AF843743

History of Living Things
Published in the British Isles and Europe by
John Murray (Publishers) Ltd.,
50 Albemarle Street, London W1.
ISBN 0 7195 3028 8

Developed at the Centre for the Development of Learning Materials
Macquarie House, Church Street, Ryde, N.S.W.

Prepared as part of the biology section of the
National Science Curriculum Materials Project

NATIONAL SCIENCE CURRICULUM MATERIALS PROJECT

B8

History of Living Things

Nan Kelly,
Sydney Teachers' College

The Jacaranda Press

First published 1971 by
JACARANDA PRESS PTY LTD
46 Douglas Street, Milton, Q.
142 Victoria Road, Marrickville, N.S.W.
162 Albert Road, South Melbourne, Vic.
142 Colin Street, West Perth, W.A.
298A Grange Road, Flinders Park, S.A.
57 France Street, Auckland, N.Z.
P.O. Box 3395, Port Moresby, T.P.N.G.

Positives supplied by
Press Etching Pty Ltd, Brisbane
Printed in Hong Kong

National Library of Australia
Card Number and ISBN 0 7016 0374 7

Contents

We are indebted to:
The British High Commission (Sydney Office); the Consulate-General of Israel; Mrs Lesley Maynard, Department of Anthropology, Sydney University, and Mr A. G. Thorne, Anatomy Department, University of Sydney, for photographs included in figure 1;
Trustees of the Australian Museum, Sydney, for photographs in figure 6;
The Geological and Mining Museum, Sydney, for the photographs in figure 7;
The American Museum of Natural History, New York, for the photographs in figures 10, 11, 27, 30;
Dr L. H. Robbins, Brisbane, for the photographs in figure 17; and
The Embassy of the USSR for the photograph in figure 28.

After completing this booklet you should know:

- whether it is possible to write a history of living things;
- hypotheses concerning the origin and development of life on earth;
- how fossils show that forms of life existed in prehistoric times, but that the record is incomplete;
- that the relative ages of fossils can be determined, and that this provides evidence of organisms and ecosystems changing through time; and
- evidence indicating that life has developed progressively over millions of years in a process of evolution.

You should be able to:

- identify and discuss the nature of fossils representing various periods in life history;
- appreciate that science may advance when the results of more than one line of research are collated;
- accept the fact that the biologist may need to draw references from incomplete evidence;
- realize that many biological problems are still unsolved; and
- speculate on the possible origin of components of the universe.

How Human History is Written

History is what happens through a span of time. A scientific method, an organised way of thinking about a problem already defined, is needed for writing history. Historians of human events usually follow a research procedure, and employ at least four major processes:

(**1**) They compile chronicles of events from reports written at the time. (Johannes Nohl, a modern German historian, did this for the Great Plague of 1644. He states: 'The *purpose* of my work is, in the first instance, to demonstrate the devastating inroads made in former times by epidemics, into the spiritual and social life of the various nations'.)

(**2**) They collect data from primary and secondary sources, and organise this material.

(**3**) They attempt to evaluate the data collected, critically examining it for authenticity and answers to the problem.

(**4**) And finally, they present the facts in a readable form which involves exposition, extrapolation and interpretation. (Nohl states: 'This book is an attempt to depict the plague in its entirety with all its social consequences'.)

Now, in most cases, history differs from science in that it is not based on experimentation, but upon reports of events that cannot be exactly repeated. However, similar events may occur or be designed to occur, and such a re-creation of an historical event made world news in 1969. Thor Heyerdahl, the Norwegian explorer, designed and sailed in a papyrus boat such as the early Egyptians used. A news report stated:

> 'Heyerdahl is trying to show that Atlantic currents could have carried ancient Egyptian sailors to American shores in pre-Columbian times, leaving cultural traces in such civilizations as the Olmecs; and he faithfully followed Pharaonic traditions in designing and provisioning a papyrus boat he called *Ra*'. (*Sydney Sunday Telegraph,* 13th July, 1969).

We shall see that experimenters in the field of the origin of life are attempting to re-create historical fact, too, by working with what they think are the ingredients of the earth's primeval atmosphere. At the same time, it is interesting to think that life is being developed, day by day, in an immense variety of living organisms.

Modern historians use written sources, such as chronicles or memoirs and diaries; oral tradition, such as tales and sagas (some of which have formed the lyrics for modern music); artistic reproductions, such as sculptures, coins and monuments; and relics, such as human remains, implements and inscriptions. (See figure 1.)

In thinking about the history of the earth, we not only have counterparts of these sources to investigate, but life in all its variety abounds around us. With careful reconstruction of earlier environments and precise observations of modern forms of life, it is possible to propose hypotheses about the origin and development of life on earth. Presenting such hypotheses and modifying them in the light of increasing knowledge makes it possible to write about the history of life on earth.

What resources do scientists have for investigating past events on the surface of this Earth? They include the following:

- The growing knowledge, and development of techniques, that enable scientists to tell us more about the age of the Earth's crust and the stages in its formation. Nature is being forced to yield its secrets at an ever-increasing rate.
- The continuing discovery and investigation of 'relics' of life on Earth—fossils and fossil remains.
- The opportunity to collate many of the findings in different areas of scientific investigation, to draw inferences from incomplete evidence and to synthesize ideas from more than one line of research. (Can you recall Mendel's hypotheses, and the results of modern research that so many years later supported them?)
- The opportunities to make continual observations of life on earth today, which in many cases mirrors the early history of life.

However, uncertainty still remains—must any 'history of life on Earth' be only an hypothesis, partially tested? Perhaps you will be able to answer this question when the evidence has been presented.

Time is Measured by Events

A basic requirement for our investigation is a **time scale** on which to place events. But to determine where our time scale should start we must ask whether there was a beginning of time itself. That is a question which science at present seems unable to answer.

Figure 1. Materials and resources used in recording Human History.
Stonehenge: Why is this symbolic of man's attempt to measure and understand time?
Coin: How can the study of coinage help us to understand history?
Script: Why has the study of scripts been important in understanding the history of man at different times and in different places?
Painting: What could aboriginal paintings tell us of history?
Excavations: How can they be dated?

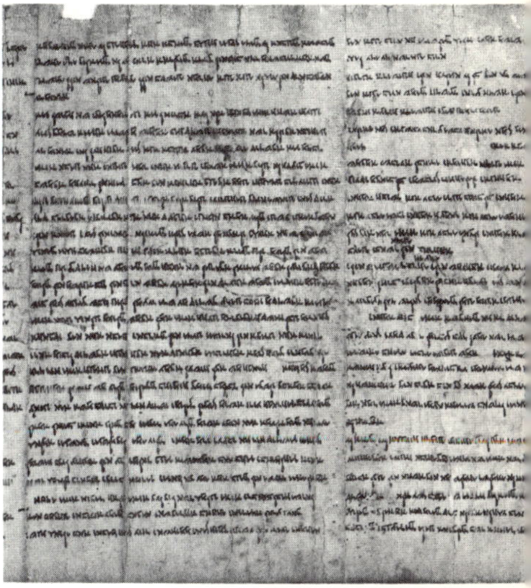

The Geological Time Scale

ERA	PERIOD	EPOCH	MILLIONS OF YEARS AGO (APPROX.)	DURATION IN MILLIONS OF YEARS (APPROX.)
CAINOZOIC	QUATERNARY	RECENT, PLEISTOCENE	0–1	1
	TERTIARY	PLIOCENE	1–13	13
		MIOCENE	13–25	12
		OLIGOCENE	25–36	11
		EOCENE	36–58	6
		PALAEOCENE	58–63	
MESOZOIC	CRETACEOUS		63–135	72
	JURASSIC		135–181	46
	TRIASSIC		181–230	49
PALAEOZOIC	PERMIAN		230–280	50
	PENNSYLVANIAN		280–310	30
	MISSISSIPPIAN		310–345	35
	DEVONIAN		345–405	60
	SILURIAN		405–425	20
	ORDOVICIAN		425–500	75
	CAMBRIAN		500–600	100
PRECAMBRIAN	UPPER, MIDDLE, LOWER	Although many local subdivisions are recognized, no world-wide system has been evolved. The Precambrian lasted for at least 2½ billion years. Oldest dated rocks are at least 2700 million, possibly 3300 million, years old.		

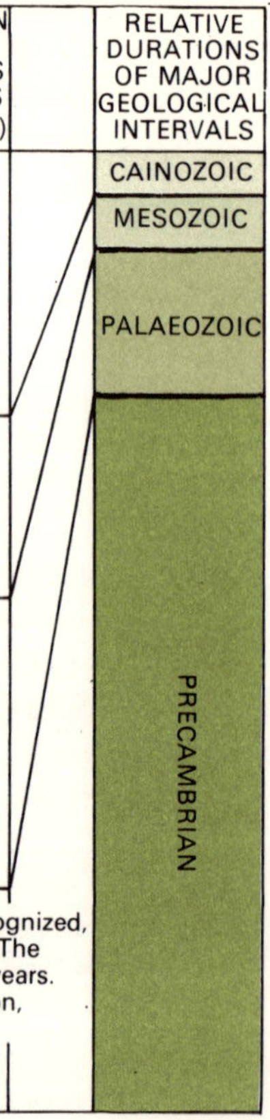

Figure 2. The Geologic Time Scale. Express the relative durations of major geological intervals as a ratio, one to another.

The present-day geological time scale has been developed over a period of more than 200 years. One of the first attempts at working out such a scale was made by Johann Lehmann (1719-67) who developed a relative time scale on the basis of rocks exposed in central Europe where he lived. His classification can be briefly described as follows;

Youngest: loosely consolidated sands and gravels;

Intermediate: secondary, layered rocks containing the preserved remains of plants and animals and particles eroded from older rocks;

Oldest: unlayered rocks which did not contain the remains of plants and animals and believed to be of chemical origin prior to the beginning of life.

Real progress in organising the rock record was not made until late into the eighteenth century, when the basic theories about the relative dating of rocks were formulated. These theories were proposed by various people, particularly Nicholas Steno, James Hutton and William Smith. A more detailed account is given in the NSCM Booklet G15 Geological Time.

Layers deposited first are covered by layers deposited later, therefore the oldest bed is on the bottom. This relationship between the youngest and oldest rocks is called **the principle of superposition.**

This seems so elementary and obvious to us now that we wonder why it was not thought of and applied earlier in history. Like other principles such as Darwin's suggestions for the origin of life on Earth, it took a long time to be recognised.

James Hutton's observations led him to the conclusion that the processes affecting the earth today also affected the Earth in the past **the principle of the uniformity of process.**

William Smith observed that fossils differed, from rock layer to rock layer, in a sequence of sedimentary rocks, and that different layers could be identified by the fossils they contained. This idea is called **fossil correlation.**

Perhaps the converse is also true, that different fossils could be identified by the different layers in which they were found. Do you think so?

The next step was to calibrate the time scale, that is, to assign some length of time to each major section. This was made possible by using radioactive substances that decay at measured rates, and hence, when discovered in the Earth's crust, can be identified and dated. The **radioactive dating method** that has been most widely used, depends on the disintegration of an isotope of the element uranium, called **uranium-238** (U^{238}). This disintegrates through a series of fourteen steps, finally producing lead-206 (Pb^{206}), a stable isotope that undergoes no further change. The half-life, that is the length of time for half of the atomic nuclei in any given quantity of the material to decay or break down, is incredibly long, namely, 4510 million years (4.51×10^9). Some intrusive igneous rocks in Canada are about 3.3×10^9 years old. Could these ancient rocks be part of the original crust of the Earth?

One isotope used widely in dating events of relatively recent time is **carbon-14**, with a half-life of 5568 years. Once carbon-14 atoms are formed, they readily unite with oxygen to form carbon dioxide. This reaction forms part of the carbon cycle in nature and you may be able to speculate how this study could help us in dating organic remains.

It is apparent that, as a clock face is a reference scale for a twenty-four-hour day, so the geological time scale is a reference scale for Earth history. It broadly subdivides geologic history into workable units. In the time scale, the Earth's history is divided into units, and scientists have established this table for convenience in discussing the rock and fossil record in a time sequence. The dividing lines between the units of eras, periods, and epochs correspond to changes of some kind, such as changes in life forms or mountain building. Two examples are: the extinction of dinosaurs dividing the Mesozoic from the Cainozoic era; and the uplift of the Appalachian geosyncline, which took place at the end of the Palaeozoic era. In no known case is the dividing line a sharp one. It is more a zone of transition, an interface of time.

In comparison with the age of the Earth, the span of man's existence is like the blink of an eye. It is interesting to think that this so newly evolved animal is the only one able to compile a geological time scale and use it. How is this so?

Origin of Life

A History of Ideas

Man has long speculated about the origin of life and the origin of the several million different kinds of living things, that is, the origin of new species of different organisms. But these are distinct problems and the problem of the origin of life involves the problem of the first organisms. The cell is the link.

For the Babylonians and ancient Egyptians, the answer to the origin of life, was **the theory of spontaneous generation.** After all, they thought that worms, toads, eels and mice formed from the silt of the Nile river. The ancient Chinese observed aphids on bamboo shoots in humid weather and concluded that they arose spontaneously. The famous Greek philosopher, Aristotle, formulated an hypothesis: Some kinds of matter, he suggested, contained an 'active principle', and each was different in different eggs. Just how untenable is that hypothesis even today? The scientists of the seventeenth century generally held spontaneous generation as dogma until such men as Francesco Redi, Spallanzani, and finally Louis Pasteur refuted it by controlled experimentation. One of biology's most important generalizations **life arises only from pre-existing life** cleared the way for the most vexing question of all:

Has there always been life on Earth, or did life have a beginning; and if a beginning, how and where did life first arise?

As historians, it is necessary for us to seek some relics, descriptions and explanations of early life and life forms. As scientists, we are faced with the need to investigate any recent experimentation and hypotheses explaining the origin of life on earth.

No one really knows what things were like four thousand million years ago, and without the invention of a time machine presumably no one ever will. Although some clues have resulted from information brought back by Apollo flights XI to XIV, at the best, only intelligent guesses, or hypotheses, can be proposed, and then tested, by trying to re-create the environment and the conditions of primitive earth on the laboratory bench. Information gleaned from studies in astronomy, astrophysics, chemistry, geology, specific meteoritic studies, biochemistry and biology has led to the belief that the commonest chemicals present in the primeval atmosphere were methane, ammonia, water and nitrogen.

Reviewing the Research

Inquiry and experimentation commenced in the early 1950s, sparked off by the speculations of a Russian scientist, A. I. Oparin, who suggested that the organic compounds which serve as a basis of life, were formed when the earth had such an atmosphere. He suggested that these relatively simple substances gradually combined into increasingly complex molecules, and were finally self-producing.

In May, 1953 the following precise and simple account, by S. M. Miller, appeared in the research journal *Science:*

'The idea that organic compounds which serve as the basis of life were formed when the Earth had an atmosphere of methane, ammonia, water and hydrogen instead of carbon dioxide, nitrogen, oxygen and water, was suggested by Oparin, Urey and Benal. In order to test this hypothesis, an apparatus was built to circulate CH_4, NH_3, and H_2 past an electric discharge. The resulting mixture has been tested for amino acids by paper chromatography ... Electrical discharge may have played a significant role in the formation of compounds in the primitive atmosphere ... The acids and amino acids formed in the discharge, not being volatile, accumulated in the water phase ... **The amino acids are not due to living organisms** because the growth would be prevented by the boiling H_2O during the run and by the $HgCl_2$ $Ba(OH)_2$, H_2SO_4, during the analysis ... glycine, α-alanine, and β-alanine are identified ... In the apparatus an attempt was made to duplicate a primitive atmosphere of the Earth and to obtain the optimum conditions for the formation of amino acids ... It is possible that, with more efficient apparatus, this type of process would be a way of commercially producing amino acids'. *Science* vol. 117, 15 May, 1953.

This report illustrates several of the points associated with excellence in scientific research—an hypothesis to support, modify or

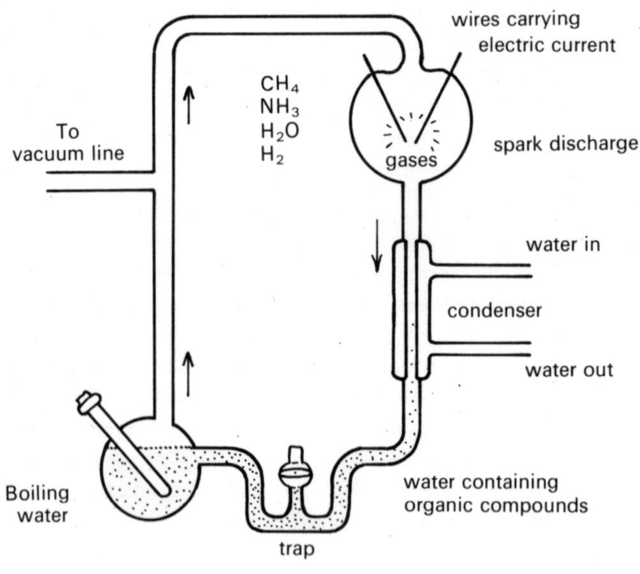

Figure 3. A diagram of the Miller spark discharge apparatus

discard; a controlled experiment; initiative in the design of apparatus; and another proposal with a problem built into it, to stimulate further research. It is no wonder that this short, simple report set in train a tremendous amount of inquiry and experimentation, as man sought the nature of his origins.

Emergent Facts

The analysis of the solution in Miller's discharge tube showed the presence of amino acids, the structural units of proteins. Doctor Sydney Fox of Florida State University added to the story by taking a mixture of eighteen to twenty amino acids and heating it. On cooling, many of the amino acids had bonded together to form chains, not unlike those characteristic of protein molecules.

Other scientists, such as the geochemist Melvin Calvin, of the University of California, using different kinds of radiation sources, have since synthesized purine and pyrimidene bases (the complex molecule forming the 'rungs' of the DNA molecule model) and even ATP (adenosine triphosphate) and simple nucleic acids from carbon dioxide, water and ammonia.

It seems reasonable to argue that the development of **heterotrophs** (dependent feeders) on the surface of this Earth would depend on a supply of complex organic molecules such as purines, sugars, amino acids, nucleotides, polypeptides and ATP, and that these could have been formed from simple molecules of NH_2, H_2O, H_2 and CH_4 in the primitive atmosphere.

But what events occurred before **non-living matter**, consisting of the basic organic molecules for life, became **living matter**? It is all very well doing experiments with simulated 'broths' to re-create primordial conditions under which life began. But how far back down the geological time scale can you chase out what really happened? Fossils of what were apparently living forms have been found in rocks that are up to some 2700 million years old. But beyond that, what? The goal of Calvin's inquiry is to discover, if possible, the interface between the purely non-living, chemical part of evolution and that which began when life, in the form of self-replicating objects, appeared. In 1968, Calvin's worry was whether he could tell a living hydrocarbon spectrum—used for distinguishing molecular architecture—from a non-living one.

An Hypothesis Regarding the Origin of Life on Earth

On reviewing these research reports, some thought will lead us to suggest that several steps must have occurred.

- A controllable source of energy is always necessary for chemical reactions. In the primitive atmosphere, would this have been ultra-violet light, electrical discharge or (as Professor Elpiner at the Biophysics Institute in Moscow has shown) could it be ultrasonic energy? In a series of experiments, Elpiner succeeded in synthesizing several amino acids, using ultrasonic waves as a source of energy.
- The next step would be the synthesis of organic molecules from inorganic molecules—amino acids from simpler molecules. The scientists discussed above have shown that this can be done.
- These new molecules would need to be built into structures that provide a basis for higher levels of organisation. But we seem a long way from cells—'the units of structure and function of living things'. We have to hypothesize that an aggregation of complex organic molecules into droplets, possibly similar to **coacervates** occurred. Oparin suggests that coacervates are simply isolated parts of the primeval 'broth', separated from the solution by surface boundaries, but able to interact with the environment.

	Raw material	Energy Source	End Product	Time, in millions of years
BIO LOGICAL PERIOD	Other organisms	+ Respiration of atmospheric oxygen →	Animals	600
	Atmospheric CO_2	+ Solar radiation →	Photosynthetic plants	2000
	Organic soup	+ Fermentation →	Primitive non-photo synthetic organisms	3500
PRE-BIO LOGICAL PERIOD	Water, atmospheric ammonia and methane	+ Solar radiation and electrical discharges, also ultrasonic waves →	Organic soup, of amino acids, nucleic acids, carbohydrates etc.	?

Figure 4. Stages in the development of life from non-living matter

- The next step would involve the separation of the coacervate system from the environment by a **membrane** having selective permeability properties— the 'semi-permeable membrane' of our cell studies.
- And finally, a mechanism of **self replication** must develop.

A Difficult Problem

When these steps have occurred, can this system then be regarded as living? And have we conveniently neglected the matter of **enzymes**, those biologic catalysts that are vital to the direction of chemical reactions in living matter?

In 1967, Oparin suggested that the most difficult question to answer is how enzymes came into existence. An enzyme structure is adapted in a very specific way to the particular catalytic task it must carry out in intact living systems. This appearance of adaptiveness in the simple aqueous solution of 'primeval broth' *before* the formation of such whole systems seems entirely improbable!

Did the coacervate drops, once formed, selectively absorb and concentrate small molecules from the external medium, and incorporate and change them once they were inside? Oparin has achieved this in his laboratory, and calls such drops **protobionts** . . . prototypes of life. Under the influence of wave action, could these protobionts have been broken up into daughter droplets, each with the attributes of its parents? And once this happened, would prebiological natural selection set in? Would the more 'successful' drops (those with greater catalytic ability to incorporate their surroundings) grow at the expense of the less successful? Perhaps an enormous number of alternative protein catalysts, or enzymes, must have arisen, of which only the most efficient remained. These successful proteins are probably today's **co-enzymes,** used now only as an aid, but still a vitally important aid, in an enzyme's action.

One of our problems is to remember that we are working with changes that occurred over such a vast period of time, that there is little evidence of them today. Oparin maintains that all these processes can be experimentally tested 'despite the fact that we have no *time machine'* to return and 'check them in the field'. His laboratory remains actively engaged in this field of research.

Other chemicals for which some scientists have been looking very hard are the **porphyrins,** the major constituent of chlorophyll. These molecules are capable of absorbing the energy of sunlight and converting it into energy that can be used for chemical synthesis. Perhaps you are already thinking that there is something familiar about this proposition. If your memory serves you well, you may already be thinking that porphyrins can be photosynthetic, and therefore food makers, and so providers for living things. Scientists working for the National Aeronautics Space Administration, in California, have identified microgramme quantities of porphyrins, as a result of firing a continuous electric arc through an atmosphere of methane, ammonia and water. An interesting point is that these experiments were essentially a repetition of initially unsuccessful work carried out before 1939; perhaps you could suggest why they were ultimately successful in their attempt.

Speculations on the Problem

Many scientists today continue to brew 'primeval broths'. Over the past decade just about every organic chemical found in living systems has been filtered from the sludgy mess which is usually the end-product of their experiments. These organic chemicals are: *amino acids*, which, linked together, constitute proteins; *bases and sugars* that go to make nucleic acids; *fatty acids*; *porphyrins* with a potential for absorbing radiant energy and able to be formed under non-living conditions... Quite suddenly perhaps, in prebiological times, a whole new source of energy by way of porphyrins, was available for the synthesis of organic compounds through the route of photosynthesis. The first steps had been taken along the road to the first primitive plants. This is speculation. Just finding and identifying these molecules in itself, proves absolutely nothing, but the discoveries do provide a firm foundation for further hypotheses and experimentation. This is how science works. You may find in later parts of this booklet that certain minute relics also point in this direction. This could be the pattern for the origin and evolution of life on Earth.

The Probability of Life on Other Planets in the Solar System

Man has cherished a number of illusions whilst writing his history; he originally placed his Earth at the centre of the universe, but today we must accept the observed fact that Earth is an insignificant planet, in orbit around a normal star, on the fringe of one of the innumerable galaxies. Despite this, life appears to man to be a very rare and special phenomenon in the universe. No doubt the outcome of recent and continuing space probes is going to yield more and more information, until eventually it will be possible to propose an hypothesis of the history of the Earth. Until then, we must bear certain things in mind. One of these is to conclude that the formation of a planetary system like 'ours' is by no means an exception, but quite a common occurrence in the universe. We must be quite clear about the conditions that are essential to life as we know it.

Conditions for Life

You may be able to discuss and support the following statements about the requirements of living things:

- Certain essential chemicals, mainly carbon, oxygen, hydrogen and nitrogen, seem necessary for the continuous recycling of organic molecules.
- An atmosphere is needed to moderate the temperature on the surface of the planet because it can filter harmful ultraviolet rays and excessive infra-red radiation. An atmosphere produces a green-house effect; it also provides a source of gases for the life processes of photosynthesis and respiration.
- A life-supporting planet must have gravity sufficient to hold its atmosphere.
- A life-supporting planet must be able to supply water, for without it no life, as we know it, is possible.
  (Perhaps you could list reasons why water is a necessity.)

Having stated the criteria we consider necessary, we can now, in a truly scientific manner, investigate the data that escalating space exploration is providing, and then draw our own conclusions about the possibility of life on other planets in the solar system.

Investigating Recent Data

Now if we consider the planetary system we know best, the solar system, we can study the data collated from the Soviet Venus 5 and 6 space probes, and the U.S. Mariner 5 and Apollo probes—the extent of our information in late 1969.

Table 1. The planets

Planet	Average Distance from the Sun	Diameter in miles	Relative Density	Probable Average Surface Temp. °F	Probable Atmospheric Gases
Mercury	35	3100	5.41	+600	None
Venus	66	7600	4.99	+600	CO_2-dense clouds
Earth	92	7900	5.52	+50	N_2, O_2, CO_2
Mars	141	4200	3.94	−40	N_2, CO_2, very thin
Jupiter	482	89 000	1.33	−150	H_2, He, CH_4, NH_3
Saturn	885	75 000	0.71	−250	H_2, He, CH_4 NH_3
Uranus	1782	29 000	1.56	−350	H_2, He, CH_4 NH_3
Neptune	2793	28 000	1.58	−400	H_2, He, CH_4 NH_3
Pluto	3680	3625	7.7	−400	?

What further valuable information might space probes of the future add to this data?

There are five outer planets: Jupiter, Saturn, Uranus, Neptune and Pluto: the distance between the planet and the Sun has important implications. Because they are so far from the Sun, it seems reasonable that the five outer planets are too cold for life, as we know it, to exist. What can we say of the remaining three planets, Mercury, Venus and Mars? One of these planets is obviously unsuitable for life as we know it—which one would it be?

It is indeed ironic that the planet Venus, whose orbit is closer to earth than any other, is so completely covered in featureless white clouds that the solid planetary surface is hidden from view. Radar observations suggest that the Venusian surface seems essentially desert-like, but with high temperatures, pressures and densities. Russian research dismisses Venus, justifiably, as overwhelmingly hostile.

It has been assumed in the past, that the two completely different atmospheres of Venus and Earth were derived in a quite different manner. Venus is supposed to have retained its primeval atmosphere, while earth's has been altered by biological activity. Research work suggests that the two planets evolved from identical basic constituents. The Nobel prizewinner in chemistry, H. C. Urey, has proposed that the carbon dioxide concentration of the earth might be governed by reactions such as interactions of surface calcite, quartz and magnesium silicate to give calcium silicate, magnesium silicate and CO_2 gas. Such reactions are rather slow in equilibrating at the temperatures found on Earth, and biological effects have predominated. In the case of Venusian temperatures, chemical equilibrium would be reached in a matter of days, rather than millions of years. The planets, Earth and Venus are termed true sister planets, composed of similar materials which have evolved in different ways. What then can we speculate about the possibility of life on Venus?

Recent probes of the American spacecraft, Mariner 6 and 7, passed close to Mars, in mid-1969, and now we can check the data for 'life on Mars' theories. Unlike other planets, Mars has a surface temperature in the range of that on Earth, although its average temperature is much lower. In addition, the surface of Mars is directly visible through its thin atmosphere of nitrogen and carbon dioxide, and for many years cyclic changes have been observed. But such data only pose more questions. Are the dark areas due to organic matter? Do the polar caps shrink in the Martian spring and summer, and the dark areas harden and sharpen, as though due to plants revived by the coming moisture? Are the 'canals' that early astronomers described, really mountain ridges? What can this mean in terms of environment? Do reports of quite large amounts of methane and ammonia on Mars indicate that there is life there, or that the earliest forms of life could be starting to evolve on this neighbour planet? Should this be so, how will this help us to interpret continuing discoveries of the earliest life forms on Earth?

At present, life on Mars can be considered only as a remote possibility. Its thin atmosphere, rigorous temperatures, and probably insufficient water, could support only the simplest forms of life. However, one thing we do tend to neglect—the remarkable adaptability of life we know on this Earth—where bacteria thrive without oxygen in hot springs at near-boiling temperatures and survive at temperatures below freezing point! You will realise now that science certainly does not have all the answers . . . It cannot have, without the data . . . and yet it seems that more data just pose more questions.

The fascinating problem remains, of whether Mars ever supported advanced forms of life. We have to admit that there is little evidence to suggest it. Life does not evolve rapidly, and Mars, with its escape velocity of only 3.1 miles per second, probably lost much of its atmosphere so quickly that life had not time to develop to the extent it did on Earth. The chances of finding fossil evidence on Mars are very low indeed. Yet despite all this, Mars is still unique in our experience; it is the one world known to us that may support life, even of a primitive kind.

The Probability of Life Elsewhere in the Universe

The problems of space travel within our own planetary system make it difficult to investigate the probability of life elsewhere. One would need to travel for ten years at the speed of light to reach even the nearest star that might have a planetary system. Most such stars with planetary systems are thousands of times farther away! We can only speculate on the effects we would observe if Earth had been colonised from other planets. There is some support however, for the hypothesis that life came to earth on meteorites from other parts of the universe. Two British scientists believe that they have identified, in meteorites, chemicals that can only be found in living cells. But this does not imply that there was life in the universe before Earth was formed, because meteorites are at least as old as Earth. Could it imply that such chemicals were carried in the original meteorite as it plummeted to Earth? The view that the 'broth' of organic chemicals, evolved on Earth by natural processes, should have left traces in the most ancient rocks, is not supported by investigations of these rocks. So perhaps this suggests that a meteorite may have seeded the primitive world with life. There is a chemical in certain meteorites, forming 4 per cent of the meteorite's weight, which is believed to be **sporopollenin**, and this can only be produced inside living cells!

What do the great scientists think? The American astronomer, Shapley, recently used a statistical method to arrive at the minimal figure for the frequency of life-bearing planets. It gives the remarkable result that in the universe at least 100 million planets offer possibilities favourable to the development of life. The English astronomer, Fred Hoyle, suggests that there are a million planets in our own

Milky Way where physico-chemical conditions are the same as Earth. The Swedish scientist, Svante Arrhenius, fifty years ago worked out a plausible explanation of life on this planet. He suggested that certain spores could have been whirled up into the higher layers of the atmosphere of a planet such as ours, and could have been widely distributed in outer space, ready to be swept up by some other planet with suitable nutrients. It is not difficult to think of objections to this theory when we consider the problems encountered by the space probes of our day.

There is a possibility that life somewhere in the universe has evolved a human or superhuman intelligence that could communicate across space by transmitting meaningful patterns of radio signals. Several large radiotelescopes on Earth are now searching. It would be a dramatic achievement if life could be discovered elsewhere in the Solar System, by scientists recognising these signals.

Back on Earth we can only postulate, with certain probability, that the surface of the Earth 5000 million years ago was composed mainly of inorganic compounds. Three million or four million years later we suddenly find fully-developed organisms floating in the seas, slowly crawling in lagoons along the ancient shores, searching, pulsating with life. Is the contrast too great? How did it come about?

A discussion of life on Earth and organisms in the Precambrian period, about 600 million years before our time, is our starting point for Earth history. We can hope to reconstruct the transformation of organic carbon material through the ages that passed after the planet Earth was formed, until we see organisms in the Cambrian seas. On and on, through the great eras of time, the Palaeozoic, Mesozoic, and Cainozoic, until we see that man is walking on two legs. He has a brain in his head, two eyes in the same plane, and a height of over five feet. We are *Homo sapiens,* products of Earth history.

The Origin and Significance of Fossils

Organizing the Record of Life on Earth

You will recall that the historian bases much of his interpretation on various types of relics such as human remains, utensils, weapons, inscriptions and monuments. Such relics or remains of animals and plants, preserved in the surface of the Earth, are called **fossils**. The term is derived from the Latin root *fossilis*—to dig. The study of fossils (and, thereby, the history of life on Earth) is called **palaeontology**. (Greek; *palaios*—ancient.) You may find it useful to look back to figure 2 and copy the geological time scale onto a firm piece of cardboard to use as a bookmark for ready reference in later reading.

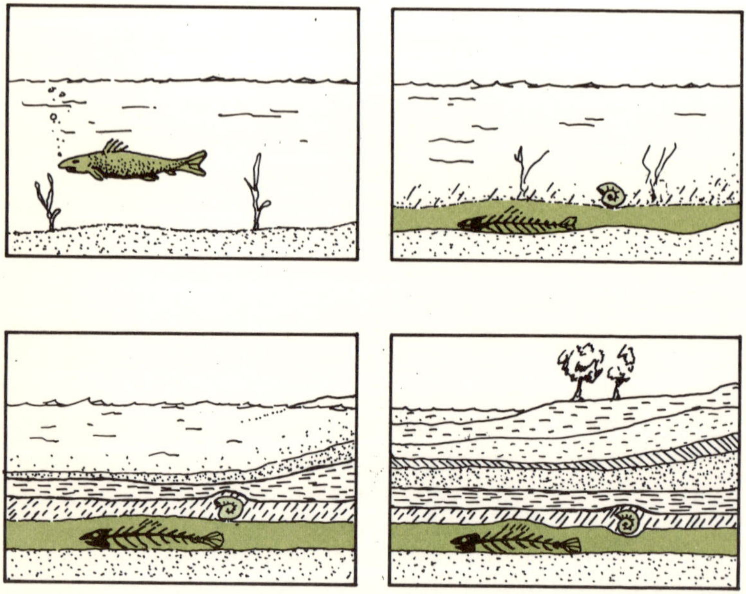

Figure 5. How fossils are preserved in rocks. Would it be likely for a mammalian fossil to be found with the other fossils in the example given above?

Not all creatures become fossils after they die. They have to be buried in a certain way. Even then, usually only their hard parts, such as shells, bones, wood and teeth are preserved. The ways in which fossils form are discussed in some detail in the NSCM Booklet G14 Fossils.

In the hypothesis proposed in the first section about the origin of life on Earth, the importance of molecules becoming more complex, and thereby gaining biological significance, was stressed. It it were possible to find evidence of such molecules in dated rocks, we would have evidence to support the hypothesis. Organic chemists and geologists have joined forces in the search for **chemical fossils**. They are

Table 2. Chemical fossils

Organic compounds originally synthesized by living organisms and more or less modified have now been found in many ancient rocks that began as sediments. The dark bars indicate reasonably reliable identification; the light bars, unconfirmed reports.

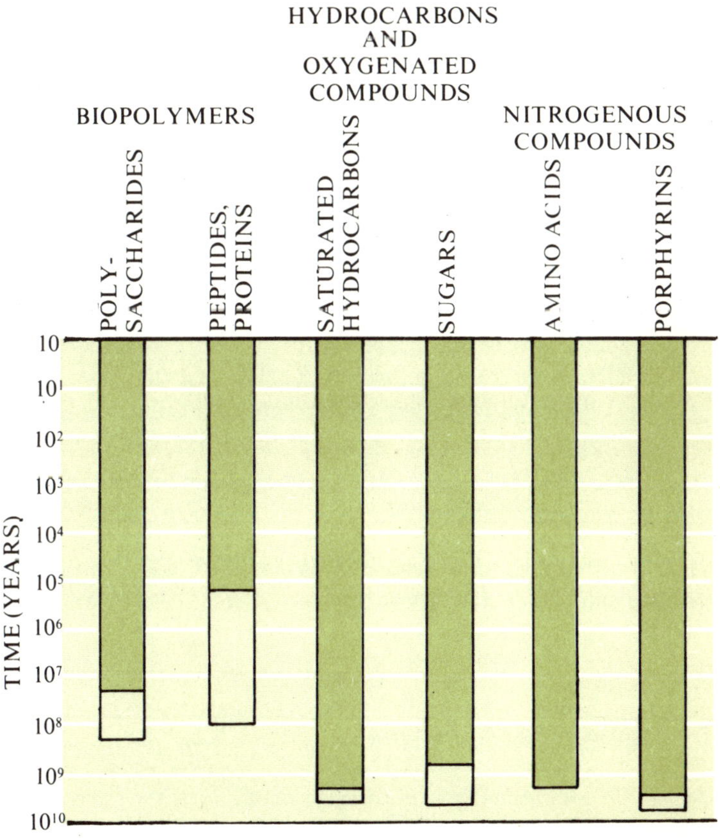

Revise your knowledge of the molecular structure of these compounds. What generalized statement about these compounds are you now in a position to make?

organic molecules, that have survived unchanged or little altered from the time when they were part of organisms that have long since vanished. This search does not require the presence of the usual kind of fossil—a shape or an actual hard form in the rock. The fossil molecules can be extracted and identified even when the organism has completely disintegrated and the organic molecules have diffused into the surrounding material. In fact, the term **biological marker** is now being applied to organic substances that show pronounced resistance to chemical change, and whose molecular structure shows that they have been formed in significant amounts, by biological processes only.

Hydrocarbons, which are comparatively resistant to chemical and biological change, resemble the hard parts of organisms. Whereas, biologically important molecules such as nucleic acids, proteins and polysaccharides decompose readily. However, several workers, such as the American Philip Abelson, have found protein in well-protected sites, such as between the sheets of crystals in fossil shells and bones. From chemical fossils it may be possible to infer the biological class, or even species, of the organisms that gave rise to them. It is uncertain however, whether the biochemistry of ancient organisms was generally the same as the biochemistry of the present day organisms. Finally, there is little known of the chemical changes wrought in organic substances when they are entombed for long periods in the rock of a fossil matrix. The records of fossil molecules that have been extracted and identified shown in table 2, are the work of Eglinton and Calvin, and were reported in *Scientific American* in 1967.

The better known fossil types can be classified, and many examples are available for study. A detailed account of fossil types is given in the NSCM Booklet G14 Fossils.

The Significance of Fossils

Although fossils are very interesting in themselves, they are also of great value to scientists. They can help scientists piece together something of Earth's history. Let us review some of the areas of knowledge dependent on studies of fossils.

Plants and animals have changed through time, and the change has shown a gradual and orderly progression. This is what the term

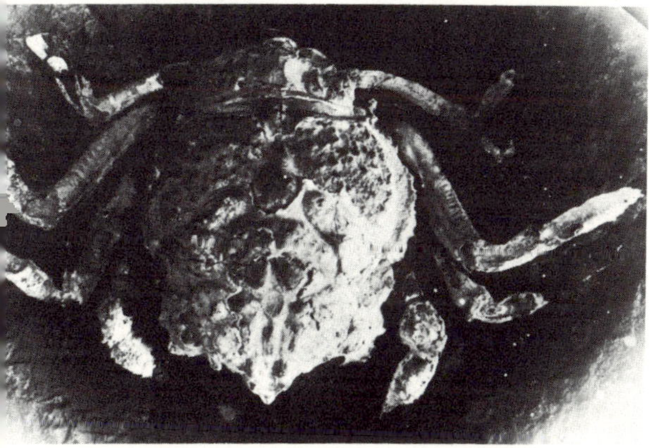

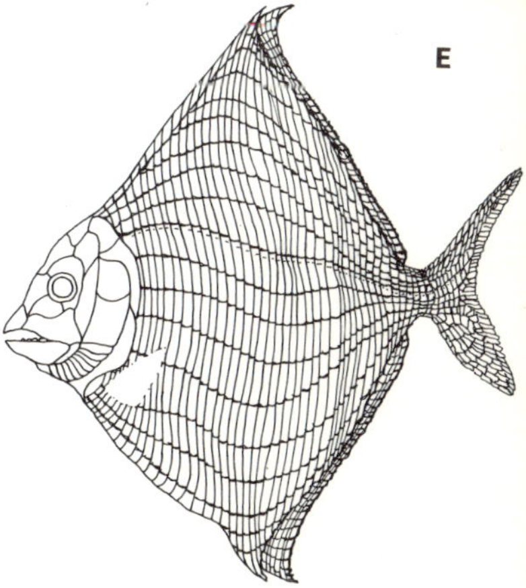

Figure 6. Some types of fossil.
(a) An insect wing.
(b) A fossil crab.
(c) A fossil leaf.
(d) Photograph of a fossil fish found in Queensland
(e) For comparison, a drawing of similar fish found in the rocks in Scotland

evolution really means—changing life. It comes from the Latin root: *volere*—to unroll. Scientists have catalogued more than a million different species of animals, and the pathways of their development can often be traced from the fossil evidence along the way.

Fossils give evidence of the distribution of former lands and seas, the existence of ancient flood plains, rivers and deserts, and of the original vegetation. Such studies are called **palaeogeography**, or ancient geography. These studies can answer questions such as:

How is it known where dinosaurs once lived?

Could the eastern part of Australia have been covered once with great swampy forests?

Fossils can tell us much about the climates and conditions of past ages. When palaeontologists look at a number of fossil organisms found in one area, they can make reasonable inferences about the nature of the ecosystem, (that is, the relationship of living things to their environment) as it existed millions of years ago. A classic example of this type of study is that of the distribution of fossil corals, and their relationship to ancient climates and seas. Modern coral reefs, such as the Great Barrier Reef, are found only in a limited range of environmental conditions. The temperatures are not less than 25° C; the clear salt water is up to 200 feet (60 metres) deep, exposed to tides, with waves and currents to bring adequate supplies of food and oxygen. Coral reefs are found in the warm current regions of the world, that is, between latitudes 30° N, and 25° S. But fossil reef deposits are found in many parts of the world where the present conditions do not apply.

Is it reasonable to assume:

(**a**) that at a particular period of geological time, the climate and sea conditions where fossil reefs occur must have closely resembled those off the coast of Queensland today, or

(**b**) that the organisms forming the fossil reefs, lived and colonised under different conditions?

Such inquiry leaves us with the problem:

Can we be sure that the distant ancestors of a warm water animal or plant was strictly a warm water organism?

Fossils are clues to Earth history because one of the most important uses of fossils is *correlating* rocks in one area with another.

Some fossils represent plants and animals that survived only a short time in geological history but which were widely distributed.

Some of these fossils are so characteristic of certain geological times that they have been called *guide or marker fossils* and are specially useful in identifying the rock layers that contain them. For example, when dinosaur remains are found, it is known that the rocks are Mesozoic in age. Why?

Certain groups of rocks overlie each other, and their stratigraphical order—the geological column—gives clues, through fossil evidence, to their relationships and ages.

Figure 7. A useful guide fossil. The cylindrical piece of rock is a section of a core taken from an oil-well at a depth of over 7000 feet. The smaller specimen was collected on the surface about 100 miles away. What do the specimens tell about the surface rocks and deeply buried beds?

Man daily probes the Earth's crust for the fossil fuels he requires (as can be seen in Australian mining and stock market reports) as geologists bore for coal deposits, natural gas and petroleum. It should not be too difficult for you to find an actual reference to recent successful drilling that may have been due to the study of fossil plants and animals. Some genera live for enormous periods of time. Veterans in the struggle for existence are the most ancient known genus alive, the brachiopod, *Lingula*. It has survived through a life history of innumerable generations spread over at least 3.5 x 10 (350 million) years. The **crossopterygians**, rarest of all the bony fish, were thought to have been extinct about 75 million years ago, until one was discovered alive off South Africa in 1939—the find of the century! The tuatura of New Zealand, dating from the Triassic period, and the lung fish found in Queensland river waters, are also very long-lived types, commonly referred to as *living fossils*.

Although the fossil record is admittedly incomplete, nevertheless, fossils are facts, and the accumulation of new evidence is progressing so rapidly through the work of palaeontologists that the broad outlines of past biological history are firmly established. Information supplied from geology, geochemistry, biochemistry, microbiology and micropalaeontology and the atmosphere sciences, is making it possible to trace the origin of plant life on Earth, 3 x 10^9 years back into the Precambrian period.

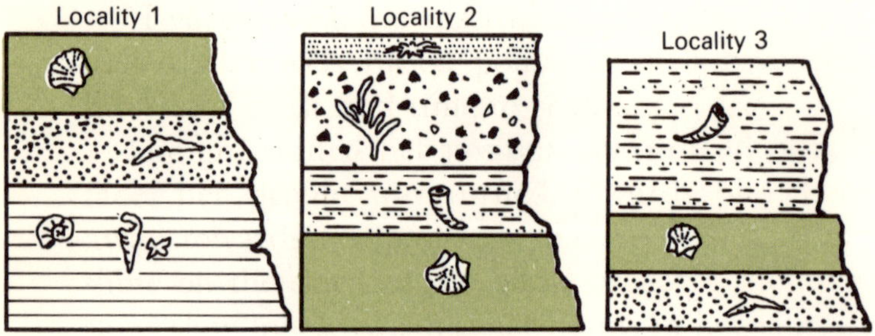

Figure 8. Sequences of fossil-bearing rocks in three localities. Relative ages of fossils can be estimated by comparing the fossils in one layer with those in another. In the case shown, which locality bears the oldest fossils and in which bed do they occur?
Which is the youngest bed, and in which locality is it?
Palaeontologists call this method correlation.

Overview of History of Life on Earth

Collecting the Data

We have seen, in the previous chapters, that fossils provide a basis for reconstructing past ecosystems. Even microfossils are helping to yield the secrets of the Earth from Precambrian times, which lasted from 5000 million years ago until 600 million years ago—an enormous span of time.

Before we begin to look at the record of life preserved in the rocks, we have to gain some perspective about the age of the Earth. According to latest estimates, the Earth is approximately 5000 million or 5 x 10^9 years old. We have gradually become conscious of the fact that our own evolution, from advanced primates, to *Homo sapiens*, represents a minimal fraction of the time that has passed since the formation of the Earth from cosmic dust. If we imagine that this span of time is represented by one year, we find that:

> 'from January 1st to December 1st there is little to be noted concerning organisms as we define them. By the middle of December, we can ascertain that life exists in the form of quite advanced organisms in the Cambrian seas. The latter part of the last week of the year finds dinosaurs in full activity, and the prehistoric bird, *Archaeopteryx*, flying with difficulty between primitive pine trees. During the last two days of the year some newcomers appear on the scene, some mammals, the sabre-toothed tiger representing their uninhibited vitality. During the last day, one hour before midnight, the first manlike creatures appear: *Sinanthropus*, *Pithecanthropus*, and their African colleagues. The last minutes represent the last glacial epoch. The last second of the year contains all the latest news—the fall of the Bastille, the battle of Trafalgar, Darwin's Origins of Species, two world wars and minor wars, the vacuum cleaner and nuclear fission!' (Ehrensvard—*Life—Origin & Development* p. 54.)

Such a picture on a broad canvas involves reconstruction from fragmentary evidence and the individual contributions of scientists who work in this field. But in any texts on Earth history, or in displays in museums, we find illustrations and dioramas (three-dimensional models) with a ring of certainty, in titles like: 'Life in Precambrian Times' or 'Evolution of the Vertebrates'. So, we are inclined to accept it all unreservedly. What information produced by specialist scientists such as geologists, chemists, biologists and palaeontologists, needs to be collated, to produce a reconstruction of the very long and vague period, from the formation of the Earth's crust, to the beginning of the Cambrian period, about 600 million years ago?

It is obviously essential to estimate the ages of the rocks and the five major rock strata accepted by **geologists**. The NSCM Booklet G15, Geological Time, discusses how the ages of strata are determined, and the problems associated with the method of dating rocks.

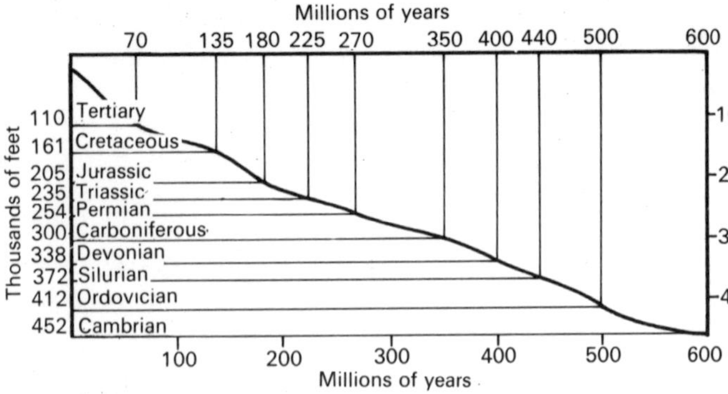

Figure 9. Ages of geological periods and thicknesses of strata

Geochemists, working with ever-increasing research techniques, have been able to tell us more about the nature of certain molecules, such as proteins and pre-proteins, occurring in ancient rocks. These fossil molecules, when extracted from rock of estimated age, can give vital leads to the development of the biochemistry of the organisms of that period. Investigation into the nature of the moon rocks brought back by the American spacecraft, Apollo XV, in 1971, will yield fascinating information about conditions on the moon.

Palaeontologists collaborate with botanists, zoologists and biochemists, working with various types of fossils—(ranging from giant skeletal bones, to microfossils)—propose life processes and patterns of behaviour from the fossils' morphology (that is, the form deducted for the organisms). So, reconstruction becomes possible.

The multifarious branches of science provide small pieces of what is little more than circumstantial evidence, when seen in isolation. When several lines of evidence come to bear on the same problem, some definitive patterns begin to develop. Collecting and collating data in this specialist age is the key to the past.

Probing the Precambrian Period

If some of the isolated facts provided by modern science are collated, something can be written about the Precambrian period. This lasted from the time of Earth's formation, until Cambrian times, a span from 5000 million years to 600 million years ago.

Reference to the geological time scale will show that Earth is estimated to be 5 x 10^9 years old. The period now known as Precambrian, lasted more than six times as long as the successive era till the present day—a short 600 million years! The oldest known and readily recognisable **microfossils** have been found in sedimentary rocks in Southern Rhodesia, and given a tentative age of 3.1 x 10^9 years. Most of the known Precambrian fossil fauna has been found in Australia. Specimens include primitive jelly fish and associated fauna, in rock of late Proterozoic age, found in 1947, at Ediacara, in the Flinders Range of South Australia. These were studied by Prof. M. Glaessner of the Adelaide University. So far, about 1500 specimens have been collected. They include: two present-day animal phyla: the **coelenterates**, including jellyfish and corals, and the **annelids**, including segmented worms. An interesting fossil find at Tea Tree Gully, near Adelaide, was thought to be a fossil arthropod, a type of ancestral sea scorpion. The remains were reconstructed and placed in a new class, but the resemblance is conjectural and not accepted—another case of uncertainty in science!
Geoffrey Eglinton, University of California, writes:

> 'Our study of chemical fossils began in 1961 when we decided to probe the sedimentary rocks of the Precambrian period in search of the earliest signs of life. This vast period of time, some 4000 million years, encompasses the beginnings of life on this planet, and its early development to the stage of organisms consisting of more than one cell. We hoped that our study would complement the efforts being made by a number of workers including one of us, (Calvin) to imitate in the laboratory the chemical evolution that must have preceded the appearance of life on Earth. We also saw the possibility that our work would be adapted to the study of meteorites and of rocks obtained from the moon or nearby planets. Thus it even includes the possibility of uncovering exotic and alien biochemistries. The exploration of the ancient rocks of the Earth provides a testing ground for the methods and the concepts involved'. (Geoffrey Eglinton and Melvin Calvin. Chemical Fossils in *Scientific American*. 1967.)

The authors of this article duly reported that, from the study of Precambrian rocks, chemical fossils reveal the presence of **phytane** and **pristane**, breakdown products of a chlorophyll molecule. Two stable carbon isotopes, C^{12} and C^{13} are present in our atmosphere in the ratio of 99:1. It is known that any biological activity that incorporates inorganic carbon (as in CO_2) into organic carbon causes

Table 3. Steps in the evolution of life

Geological time scale shows the age of some intensively studied sedimentary rocks (*left*) and the sequence of major steps in the evolution of life (*right*). The stage for biological evolution was set by chemical evolution, but the period of transition is not known.

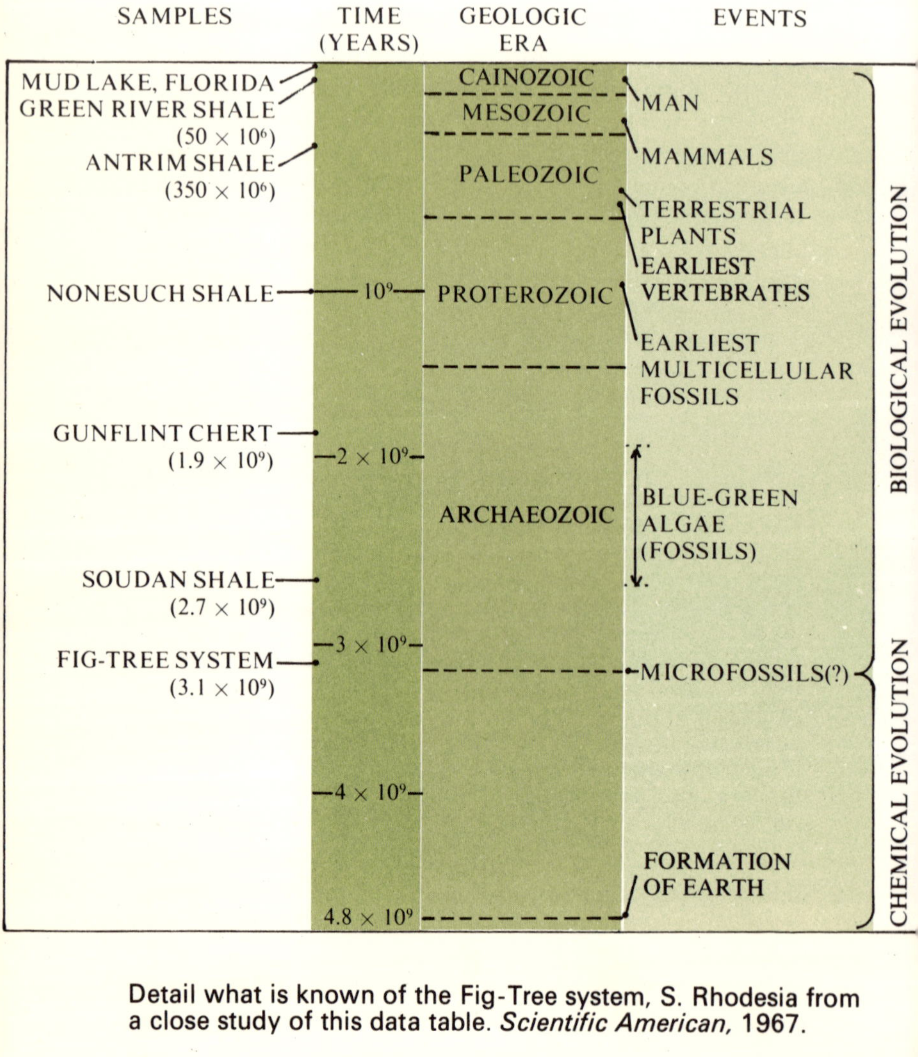

Detail what is known of the Fig-Tree system, S. Rhodesia from a close study of this data table. ***Scientific American,*** **1967.**

an enrichment of the C^{12} isotope. This would mean that organic material derived by biological activity has a slightly higher $C^{12}:C^{13}$ ratio than the parent rock. Further studies suggest that free carbon dioxide probably existed at this time. Can this biochemistry lead to the conclusion that photosynthetic organisms had evolved during this period?

Using such apparently isolated reports as these, it is possible to form some ideas about the early Precambrian period. It is then possible to work out its age, duration, the atmospheric conditions, the geological conditions, the presence of primitive organisms, and the development and relationship of such organisms. Ideas that can be tabulated, as in table 3, can be translated into definite facts, displays, written descriptions and dioramas. That is how a section of Earth history is written. However, a number of criteria must be satisfied before we are fully justified in assuming that biosynthesis occurred in Precambrian times:

- We must show that chemical fossils associated with cellular debris occurred in situ and not from contamination.
- We must also be certain that organic matter had not been produced **abiogenetically** (in the absence of life) for there is increasing evidence that many biologically significant molecules can be produced when no living organisms are present.
- We must be aware that microfossils themselves may be subject to misinterpretation.
- We must accept the probability that Precambrian microflora represent a highly selected group of organisms, and that the prevailing environments may have promoted growth and preservation of both bacteria and blue-green algae.

The use of **stromatolites** (calcareous and silicous deposits of colonial blue-green algae) in fossil correlation presents two problems.

How much of their form is true reflection of the series of alga involved?

How much is due to the plant's response to the particular environment in which it is found? A centre for extensive work is the Adelaide geosyncline and the Amadeus basin of Central Australia.

Keeping open minds about the implications of further discoveries, and the need to satisfy a number of criteria, it can be stated from the data available that:

- The Precambrian is a very long period of time during which the formation of continents may have commenced rather early, but

Table 4. Precambrian plant record

Stratigraphic deposit	Environment	Geochemistry	Metabolism	Organisms
Bitter Springs, Australia 1.0×10^9 years	Shallow marine Mats planktonic (Littoral?)	Phytane, pristane, tetrapyrolles		Procarotes and eucaryotes
Mount Isa, Australia 1.4×10^9 years	Shallow lake bottom		Aerobic respiration O_2 at 1% PAL	Bacteria(?)
Gunflint Iron Formation 1.9×10^9 years	Shallow water	Phytane, pristane, C^{13} enrichment, tetrapyrolles	Aerophillic photoautotrophs	Bacteria and blue-green algae
Fig Tree Series, S. Rhodesia 3.1×10^9 years	Aquatic (10m)	Phytane, pristane, C^{13} enrichment (C as CO_2) L—animo acids	Anaerobic Photophosphorylation(?)	Bacteria and blue-green algae
Onverwacht Series, S. Rhodesia 3.2×10^9 years	Aquatic(?)	Carbonaceous filaments	Anaerobic heterotrophs chemoorganic, chemolithic	?

Study this data table closely, and then prepare a statement about the occurrence and nature of microfossils in Australia. *New Scientist* Vol. 42. No. 648

the accumulation of continental rocks continued as a result of volcanic activity and intrusion.

- The atmosphere and water of the Earth were essentially in their present state by the end of the Precambrian.
- From a study of the microfossils, there appears to be a gradual progress from the unicellular to the filamentous, to the complex filamentous in the organisms resembling blue-green algae; there is a gradual increase in morphological complexity. This supports a theory of organic evolution.
- All the microfossils in early Precambrian times appear to have come from **procaryotic** organisms. Such organisms lack discrete nuclear membranes and do not exhibit mitosis and meiosis. Resembling these, then and now, are bacteria and the blue-green algae. The first **eucaryotic** organisms, with discrete nuclei bounded by membranes, probably appeared 1.5 x 10^9 years ago, as seen in the Bitter Springs, Australia, material, but how they actually evolved is open to speculation. Could it have been by **endosymbiosis**?

  Could a relationship between small and large procaryotic cells—have been involved?
- Bacteria-like organisms, as seen in the Fig Tree, Southern Rhodesia deposits, may have been similar to present-day photosynthetic bacteria—they do not use water as a hydrogen donor and, consequently, do not give off oxygen. The blue-green algae, although structurally different from the higher plants, evolve oxygen during photosynthesis, and therefore were the first true plants—the providers for life on Earth. Their appearance must have produced different, perhaps difficult, conditions for the anaerobic organisms. Anaerobic organisms do not require atmospheric oxygen. Nonetheless, for them, oxygen gradually became a more tolerable part of the atmosphere.

Initially, these organisms must have been marine, and living in at least 10 metres of water to prevent them from being damaged by ultraviolet light. As the oxygen concentration increased, they would have inhabited shallow water. The Bitter Springs material suggests that the primitive and minute plants were in mats or masses and even littoral, or on the edge of the land. Two atmospheric physicists, L. V. Berkner and L. C. Marshall, suggest that the dramatic expansion of animal life around 600 million years ago marks the point when free oxygen first became abundant enough to support animal respiration.

Table 5. Evolution of life and environments

ERA	PERIOD (and Epoch for Cainozoic Era)		Also approximate time since beginning	PLANTS	ANIMALS	CLIMATES
CAINOZOIC	Quaternary	Recent Pleistocene	1 000 000 years ago	Dominance of herbs Increase of herbs Decrease of trees	Human cultures Extinction of many large mammals Early man	Climatic zones and changes of season Periodic continental glaciers in north
	Tertiary	Pliocene	13 000 000 years ago	Rise of herbs Spread of grasslands	Abundant mammals	Cool and temperate climates away from equator
		Miocene	25 000 000 years ago	Development of grasses Reduction of forests	Increase of mammals	Cooling of climates
		Oligocene	36 000 000 years ago	Worldwide tropical forests	Appearance of modern mammals	Warm climates still widely spread
		Eocene	58 000 000 years ago	Extension of angio-sperms	Archaic mammals	Zoned climatic belts well established
		Palaeocene	63 000 000 years ago	Modernization of angiosperms	Evolutionary "explosion" of mammals	Developments of climatic belts
MESOZOIC	Cretaceous		135 000 000 years ago	Rise of flowering plants Decrease of conifers	Decrease of reptiles Climax of dinosaurs Beginning of teleost fishes	Birth of modern mountains Development of climatic diversity
	Jurassic		181 000 000 years ago	Conifers and cycads dominant First known flowering plants	Dominance of dinosaurs First birds Spread of reptiles	Culmination of world wide warm climates

	Triassic	230 000 000 years ago	Increase of conifers First known cycads	Transition, reptiles to mammals Rise of progressive reptiles	Worldwide tropical and subtropical climates
PALAEOZOIC	Permian	280 000 000 years ago	Dwindling of ancient plants Seed ferns extinct	First dominance of reptiles	Climates become variant Glaciation in southern hemisphere
	Carboniferous (Pennsylvanian)	310 000 000 years ago	Great tropical coal forests	Amphibians dominant on land	Uniform climates throughout world
	Carboniferous (Mississippian)	345 000 000 years ago	First known mosses, seed ferns, and conifers Early coal forests	Rise of insects	Uniform climates Spread of tropical seas
	Devonian	405 000 000 years ago	First forests First known liverworts horsetails, and ferns	First amphibians Wide radiation of early fishes	Broad distribution of uniform climates Increased temperatures
	Silurian	425 000 000 years ago	First known land plants—club mosses	Wide expansion of the invertebrates	Slight climatic cooling
	Ordovician	500 000 000 years ago	Marine algae dominant	First indication of fishes	Warm climates
	Cambrian	600 000 000 years ago	Algae, fungi, and bacteria First plant spores	Invertebrates numerous and varied	Climates become progressively warmer
PRECAMBRIAN		5 000 000 000 years ago	Bacteria, fungi, and algae	Primitive invertebrates	Cool climates at end of Precambrian Glaciation

But the next problem is to explain the short time for the development of the advanced early animals. It seems as if a longer time is needed to explain some of the early Cambrian faunas. Younger Precambrian rocks yield fossils including segmented worms, jellyfish, fragments of supposed crustaceans, and structures strongly suggestive of primitive brachiopods and sponge spicules. Now that we have some evidence of how life began, its development and diversity becomes bewildering. Most large groups of the animal and plant kingdoms are well represented in the next period, the Cambrian, but the links are vague.

Life Proceeds Through the Palaeozoic Era

This is called the **Age of Ancient Life** (Greek: *palaios* ancient + *zoo*-life) and there are a great many well preserved fossils from the 375 million years of the Palaeozoic Era. When we use the method of reconstruction described in the previous chapter, we have many more fossil types to work with. So, it is easier to extrapolate—to design an environment and pattern of life to suit the fossil and geological evidence. A total environment is made up of non-living factors, such as topography, climate, atmosphere, as well as plant and animal interaction. These are the factors to consider when reconstructing life in ancient times.

If you had a time-machine, set it for 500 million years B.C., and took off, you would have a rude shock when you arrived. If you had started from a mountain area, you might find yourself foundering in an ancient sea when you landed. The main ocean basins and continental masses probably existed in the Cambrian, but their outlines differed. Shallow seas and arms of oceans flooded right into continental blocks, giving that watery environment relished by the primitive organisms that had evolved from the primeval 'broth'. But as you looked around, you would see a scene of depressing sameness—no grass, shrubs, trees, noisy insects, birds or scurrying animals—not your type of life at all. If you wandered to the edge of the land, there would be much activity in the sea water. There you would see unrecognisable animals, floating and moving amongst the primitive plants.

The first part of this era, the **Cambrian** period, was a very long one, and the various major groups of animals enter the fossil record during its long span of some 75 to 80 million years. These history-making groups can be better met in the table which shows the number of phyla and classes of animals definitely known as fossils at the stated times, or earlier. Life was commencing to jump . . . in numbers and directions. (See tables 6 and 7.)

Table 6. Numbers of major groups of organisms surely known to have existed in sub-divisions of Cambrian and Ordovician time

	No. of Phyla	No. of Classes
CAMBRIAN Early	8	12
Middle	11	20
Late	12	22
ORDOVICIAN		
Early	12	27
Middle	12	32
Late	13	33

- How can you explain that the numbers of classes increased while the number of phyla remained constant?
- Correlating the two tables, what would be some of the 12 phyla in existence by the late Ordovician?

It is not possible here to treat the six periods of the Palaeozoic era in detail; full descriptions can be always found in other texts. It is sufficient to mention the continuing abundance of variety, and then to study, in detail, some selected fossils of which we have knowledge. Such case studies in detail can help us view the era in general.

By the end of the **Ordovician** (a term derived from the *Ordovices*, an ancient Celtic race in North Wales, where a sequence of such rocks has been extensively studied), the land was still barren and there was little life in the lakes and rivers, but the seas were full of groping, drifting organisms. One genus has survived to the present day—a brachiopod, *Lingula*, which may be representative of the oldest living multicellular thing on earth. Sponges, floating graptolites,

Table 7. The first appearance of some major groups of organisms in the fossil record. The chart is subject to revision, presenting only what is well established in 1965. Figures are in millions of years

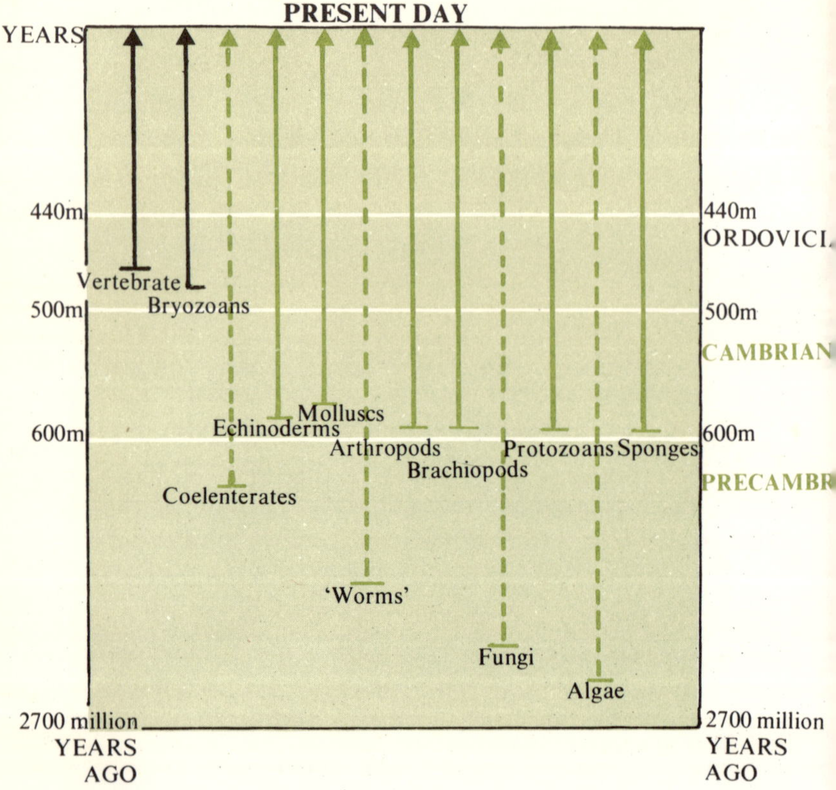

coral, bryozoans, brachiopods, pelycepods, echinoderms, sea snails, cephalopods and the most bizarre, the trilobites, lived in the Ordovician seas. (See figure 10.) In the late Ordovician, the first fossil vertebrates appear, the first fishes—jawless, primitive, few and unimpressive.

The **Silurian** (again from ancient Celtic people, the Silures), the next period, was still a time of ceaseless change in the history of life. Graptolites declined, corals expanded and the Devonian period saw accelerated evolutionary activity in many groups of animals. Fish became dominant and common, and this was known as the **Age of Fishes**. Many of the modern types began their existence at this time—the cartilaginous fishes, the bony fishes, the lung fish. The

Figure 10. Fauna of Middle Ordovician seas. Suggest some food chains that may have operated in this ecosystem

Permian and **Carboniferous** periods had much in common and can be thought of as the late Palaeozoic era.

In the Permian, changes in climate and topography sculptured the surface of the earth, and a great glaciation spread from the Antarctic to the Equator. The land became covered with low swamps and the plants grew taller to form forests over the land. Varieties of vertebrates and invertebrates colonised the new environments. Insects abounded in variety and formed the food for amphibians—clumsy brutes with four short, sprawling legs, big flattened heads, and sturdy tails. They have been named **labyrinthodonts** because of a certain tooth structure. By the beginning of the Permian, the amphibians were already outnumbered by the reptiles, and the transition to life on land was completed.

It is interesting to speculate why it appears to have been a time of crisis in the history of the marine invertebrates, and why 'land-livers' took the limelight. On land, most of the great coal forest trees became extinct, while the cycads, ginkgos and conifers spread, and soon the early flowering plants found roothold on Earth. Three hundred and seventy five million years had passed; Earth's surface had changed; and the long, slow development of organisms from water to land, and in great diversity, was under way. (See figure 11.)

Figure 11. A Carboniferous forest.
Why can this period be called Carboniferous,—carbon-bearing?
Carboniferous forests were remarkable for their large size. Why?
Insects particularly associated with flowers were absent. Can this be explained?

Some Case Studies

In human history, it is the personalities that exerted influence and left records to tell us so; these records help us put the jig-saw pieces together and fill in the details of a period. The Romans with their system of rule, the British with their concepts of justice, the French with their code for civil administration . . . a little of all these influences remains in our modern form of government. Just so, there are some organisms whose fossil evidence is reasonably complete, and it is a study of these remains that helps us to build up our ideas of life history and its evolution. The assemblages of fossils found in successive layers may be compared with the group photographs of a sporting club or team hung on the club walls. For a few years, the groups are closely similar, though a few old faces go and new ones appear. A time comes, however, when, with rare exceptions, the personnel has completely changed. So an assemblage of fossils remains but little changed throughout several adjoining layers of rock. These may then be spoken of collectively as a **zone**, and the community of fossils as the **zonal assemblage**. Zoning opened the way to many researches and discoveries, some of which threw light upon the changes undergone by the Earth's crust, others on changes in fossil remains.

B

D

E

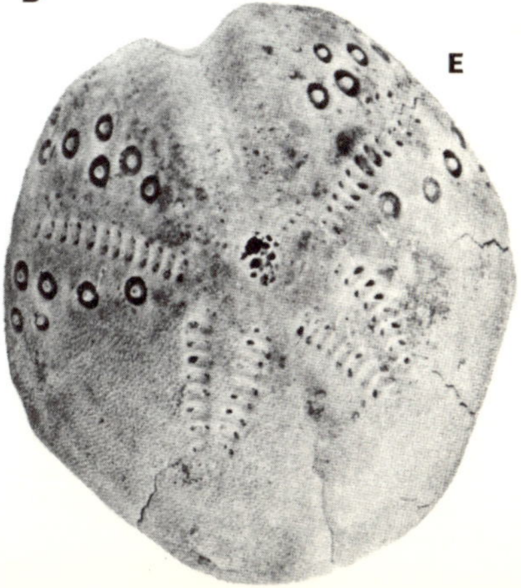

Figure 12. Organisms with complete fossil records.
(a) Foraminifera.
(b) Corals.
(c) Graptolites.
(d) Trilobites.
(e) Echinoderms.

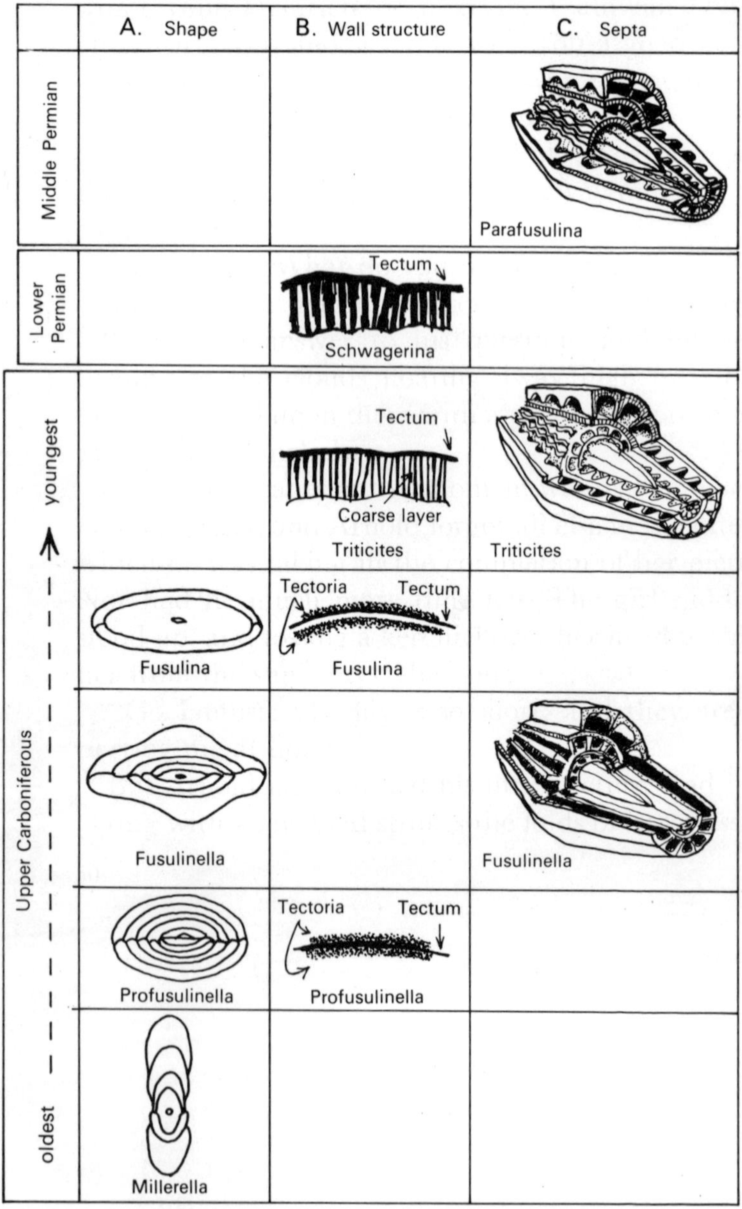

Figure 13. Evolution in the Fusilinid Foraminifera. A in the shape of the test B in the structure of the wall C in the folding of the septa.

Case Study 1. The Foraminifera or Forams

These are mainly marine members of the phylum Protozoa, or one-celled animals. You would think, perhaps, that such minute and usually soft-bodied organisms would leave little in the way of fossil evidence. Like other protozoans the forams' chief characteristic is their ability to send out protoplasmic extensions called pseudopodia. (See figure 12.) In all but the simplest forms, the protoplasm is enclosed in a shell or **test** which may be a single chamber, or may be composed of several chambers. The pseudopodia project out of one or more holes in the test, and the test may be made of sand grains cemented together (arenaceous), of chitin, or of calcite. They range in length from about 60 microns to over 250 cm. Most are bottom living, moving about on the sea floor by means of their pseudopodia. Their remains are abundant in many kinds of marine sediments, especially those of fine texture which are otherwise devoid of fossils. Correlations can be made over a wide area, because they are small, so samples of entire faunas can be collected from the cores of wells deep below the surface. This aids in the dating of unexposed rocks. Over 5000 species of forams, sometimes referred to as fusulines, have been identified.

The foraminifera offer very good evidence for evolution when a detailed study is made of their structure. A group of the larger foraminifera, the *Fusulinids*, occurs in rocks of the Upper Carboniferous and Permian, and shows the changes of structure with time.

(**a**) Their shells were spirally coiled and divided into chambers by septa which ran parallel to the axis of coiling; the last chamber opened to the exterior by a narrow slit. In the earlier forms the shell was discoidal, the smallest diameter coinciding with the axis of coiling. The shape changed progressively to become spherical, spindle-shaped and finally almost cylindrical, the largest diameter coinciding with the axis of coiling. (See figure 13.)

(**b**) The nature of the wall also shows progressive changes. In the older species it consisted of a dense central layer, the **tectum**, and secondary layers, the **tectoria**, above and below it. Later types developed a fourth layer, but the wall structure then changed markedly, and became two-layered but thicker and stouter. The tectum was retained but the tectoria were lost—the lower one being replaced by a thick coarse layer.

(**c**) The septa also show gradual change. In early forms they were flat, but in more recent species they became fluted.

It is interesting to note these gradual changes—from simple to complex form. It is more interesting to attempt to interpret them as a result of **natural selection**. There is a certain amount of opportunism in the concept of evolution—that is, to enter a new mode of life, a species must be given the opportunity to do so. Pseudopodia evidently arose first as useful supplementary feeding devices in animal flagellates. The evolutionary trends in the forams—increasing complexity of the basic shell structure, changes in wall structures and the tendency for the wall to become fluted, must all have provided opportunity for survival. You may be able to suggest what advantages these trends gave to planktonic animals.

The foraminifera are of great importance to the oil-field geologist, and widespread fossil evidence points to a widespread *geographical distribution*. The distribution of **pelagic** types, for instance, *Globigerina*, as well as the **planktonic**, is influenced largely by temperature; the former are more numerous in warm regions, the latter have their range determined by temperature and depth. The Foraminifera attain their greatest development in Tertiary and recent times. Detailed study of figure 13 will perhaps suggest reasons for their continued existence from the Palaeozoic to the present.

Case Study 2. Corals

The corals are a group of coelenterates that are found in rocks dating from the Cambrian to the present. The body is a single sac with a single opening, the mouth, surrounded by a ring of tentacles. The body wall consists of two layers of cells and is **radially symmetrical.** This type of body form is known as a **polyp** (See figure 14), in contrast to the free-swimming forms, such as jellyfish, which are known as **medusae**. In the corals and sea anemones, the body cavity is divided by radial partitions or **mesenteries**, to increase the surface area of the gut for digestion and absorption. The corals secrete a protective cup or **theca** into which the polyps can retract, and which, naturally, will leave fossil evidence of their existence. In its simplest form, the theca of a single polyp is conical or tubular, and consists of a basic disc and radial partitions, called **septa**, which alternate with the mesenteries and outer wall. The polyp lives in a cup-shaped depression at the top called a **calyx**. As the polyp grows

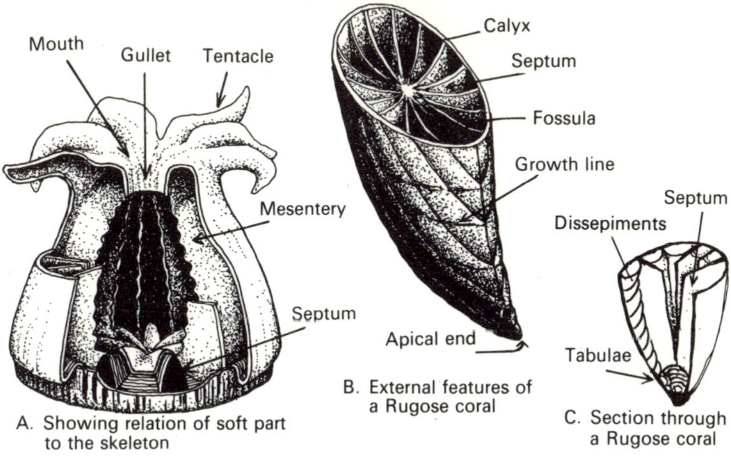

Figure 14. The structure of corals.

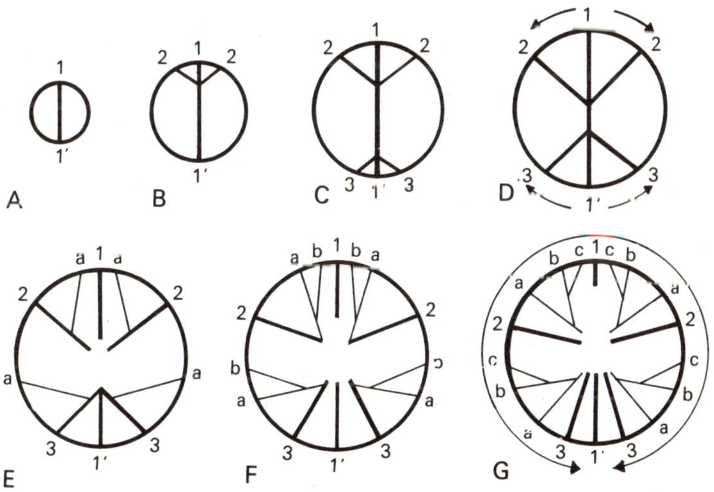

Figure 15. Development of the septa in a simple rugose coral, Zaphrentis.
1-3 primary septa
1 cardinal septum
1 counter septum
2 alar septa
3 counter-lateral septa
a, b, c, later septa (metasepta)

in size, the lower part of the theca is sealed off by transverse partitions called **tabulae**. The wall of the theca may also be thickened by small tangential plates called **dissepiments**, so that it becomes rather spongy in texture. These features, of course, give a chance of great variation.

(**a**) An interesting example of evolution has been observed in certain solitary corals that occur in Carboniferous rocks.

The particular forms belong to the genus *Zaphrentis*, in which the theca is conically shaped and slightly curved. The growth of the theca and the development of the septa can be observed by examination of a series of thin slices taken progressively higher up the theca, as shown in figure 16. Single plates give way to more septa, added in a serial fashion, to form an adult ring. Eventually, gaps appear in the septa; the largest is known as the **cardinal fossula**.

(**b**) Many specimens of *Zaphrentis* were collected from different localities and different horizons in the Carboniferous rocks of the Midland Valley of Scotland. Externally, they were all similar and were given the name *Zaphrentis delanouei*. However, when sections were cut and examined, distinct variations were found in the arrangement of the septa and in the shape of the cardinal fossula, as shown in figure 16. Different names were given to typical variations in pattern. The fossils were taken from four different horizons, often separated by hundreds of feet of unfossiliferous rock. In the second lowest horizon, the complete series, from *Z. delanouei* to *Z. disjuncta*, was present. This suggested that the various 'species'

Table 8. Percentage distribution of varieties of Zaphrentis delanouei *in Carboniferous rocks in Scotland*

Horizon	*Z.delanouei*	*Z.parallela*	*Z.constricta*	*Z.disjuncta*
Millstone grit (youngest)	—	—	5	95
Upper Limestone	—	0.3	16	83
Lower Limestone	0.7	3.3	69	28
Cementstone (oldest)	69	30	1	—

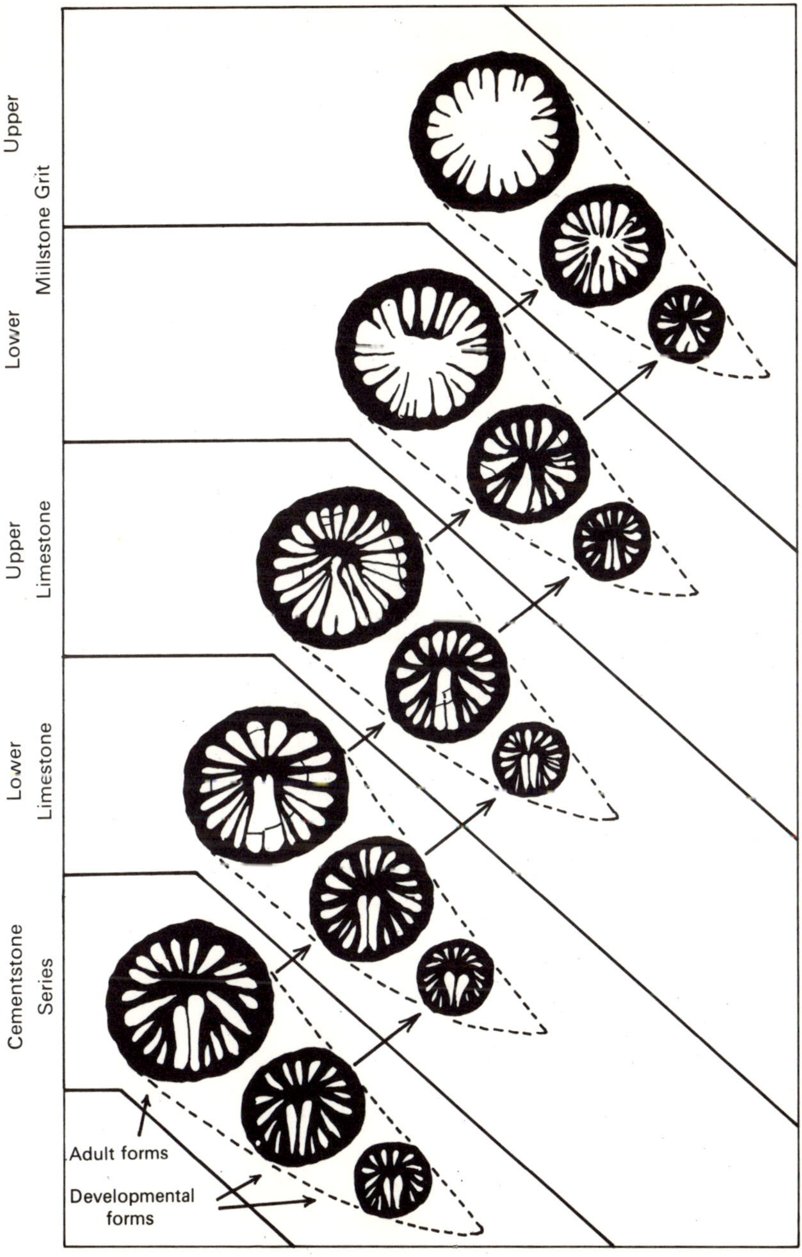

Figure 16. Five varieties of the Carboniferous rugose coral Zaphrentis, *from the Midland Valley of Scotland. Transverse sections show the arrangements of septa in the young and adult forms of each variety.*

interbred, and did not belong to different biological species, but were structural varieties of the same species. Can you suggest why this explanation is satisfactory?

In table 8, you can see that the series represented a progressive change in the community. That is one aspect of evolution.

(c) Another glance at figure 16 will show that the developmental stages of the higher varieties resemble the adult stages of the lower ones. This is an example of the **Principle of Recapitulation**, proposed by Ernest Haeckel, (1843-1919). *'Ontogeny recapitulates phylogeny'* or the development stages show signs of the characteristics of earlier adult forms. This principle has its limitations, but is often used to bridge gaps in the fossil record. Perhaps you could suggest how it is used for this purpose.

Case Study 3. Graptolites

The Graptolites are found only in Lower Palaeozoic rocks, where, owing to their abundance and to the limited range in time of both genera and species, combined with their wide geographical distribution, they are of great importance to the stratigraphical geologist (see figure 12). Floating graptolites occurred in enormous numbers and would have been the most familiar Ordovician organisms to a time-travelling zoologist. There is nothing like them today. They are usually found in shales and slates and they look like pencil-marks. The word *graptolite* means 'writing on stone'.

The graptolites were compound animals with soft parts protected by a skeleton of **chitin**. The original material of the skeleton is seldom preserved unaltered; in some cases it has been replaced by iron pyrites, but usually it has become carbonized. You may be able to recall the name given to this process of fossil formation. The soft parts of the graptolites are unknown and it is difficult to speak of their affinities with any degree of certainty. You will find further discussion of this in NSCM Booklet G14 Fossils. From comparison with living hydroids, which have a similar skeleton, we may consider it probable that each hydrotheca or cup-like structure provided protection into which the polyp could retract.

Since it does not seem as if graptolites had rootlike structures, it is not likely that they were sessile animals living on the sea floor. Furthermore, had they been rooted, some specimens would almost certainly be found in a vertical position crossing planes of lamination.

The much greater abundance of graptolites in thin, fine-grained carbonaceous shales than in thicker, coarser deposits, suggests that they lived mainly at some distance from the shore. Here, sediment was deposited in tranquil water and the carbonaceous matter in the shales may have been derived from the deposition of plant material. Is there any other explanation?

The graptolites are a group whose early forms *(Bryograptus)* appear, at first sight, to be more advanced than later types (*Monograptus*), on account of more complex branching. But in the early forms, the hydrothecae are very simple, whereas in later ones they exhibit considerable modification. You can see how selection of certain features for comparison is a very important aspect of research into evolutionary trends.

Case Study 4. Trilobites

The dominant arthropods during most of the Palaeozoic era were the now-extinct trilobites. They are perhaps one of the best-known and most easily-recognised fossil groups, and make up about 60 per cent of all Cambrian fossils. (See figure 12.)

The trilobites derive their name from the fact that the body is divided into three parts, by means of two furrows which extend from anterior to posterior extremities. This trilobation is usually conspicuous, but in a few genera is indistinct or almost obsolete. The body is oval in outline, dorsally curved, and flattened ventrally. It consists of the **head**, **thorax** and **pygidium**. They were equipped with gills for gaseous exchange, and each segment of their segmented bodies bore a pair of jointed limbs for crawling or swimming. The trilobites were abundant in the Cambrian and Ordovician periods, but were extinct by the Permian.

Examination of the trilobites from rocks of various ages shows several evolutionary trends, but not all these trends are developed *at the same rate* so that, in an individual animal, some primitive features may be present with more advanced features. The main trends are as follows:

- In the most primitive forms, the axial portion of the head or **glabellum** is narrow and only slightly raised. Later forms show an enlargement and bulging of the anterior end.
- Two evolutionary tendencies are exhibited in the eyes of trilobites—the reduction in size and eventual disappearance of eyes

in the burrowing types, and their enlargement in swimming forms.

- The thorax in primitive forms has a large number of segments, but in more advanced forms the tendency is to reduce the number of segments and for them to become wider, straighter and blunt.
- The most primitive condition of the pygidium is that of a single, rounded terminal plate. In later forms, its size is increased by the fusion of more and more segments until, in many forms, it is as large as the head. This is thought to be a protective device, as many trilobites are found rolled-up, like present day slaters, with the pygidium covering the head.

Study of such features offers a wonderful opportunity for formulating hypotheses about the trilobites' mode of life. Some forms had no eyes, and probably burrowed. Further evidence for this is that some specimens have been found containing casts of the stomach and intestines, in turn filled with stone casts, suggesting a mode of nutrition like that of an earthworm today. Other forms had greatly enlarged eyes allowing for vision downwards, suggesting a swimming mode of life. The vast enlargement of the glabellum suggests that this was modified as a floating organ, and the presence of spiny outgrowths may have assisted in the buoyancy effect. The fact that the swimming forms are found in deposits containing other planktonic organisms is further evidence for this theory.

The fossil record reveals that organisms have tended to die out simultaneously in relatively sudden world-wide *extinctions*. Apparently, in the Silurian, the environment became unfavourable for trilobites, and although they occupied several habitats and were adapted in a number of different ways, their numbers declined. The times of extinctions are often followed by periods of rapid *evolutionary radiation*, suggesting that the vacant environments left behind by extinct plants and animals provide an ideal setting for new evolutionary experimentation. The most dramatic extinctions are those near the end of the Permian and Cretaceous periods; the causes are obscure and no single change can satisfactorily account for the complex patterns of radiation and extinction found in the fossil record.

Case Study 5. Echinoderms.

The Echinoderms were another abundant and important group in late Ordovician seas, and it is not wholly unexpected that they

Figure 17. Living forms of echinoids.

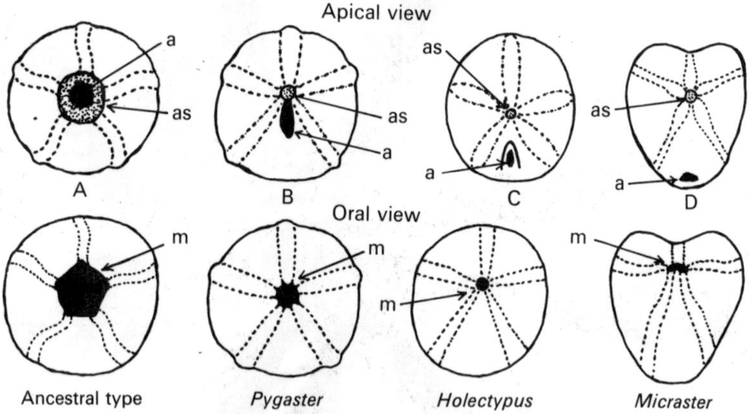

Figure 18. Diagrams to show the change from radial to bilateral symmetry in the echinoids.
(a = anus, as = apical system, m = mouth.)

should have left a fossil history because they possess an internal skeleton of calcareous plates fitted together in a flexible and rigid test. (See figure 12.) In the sea urchins, the test is hollow and globular, discoidal or heart-shaped. Living rock-dwelling sea urchins show a five-rayed symmetry, due to a double row of plates perforated by tube feet and known as **ambulacral** areas. In between are several rows of plates which bear rounded knobs for the ball and socket articulation of the spines. Echinoids showing the five-rayed symmetry are known as regular echinoids. Another group with **bilateral symmetry** has a test, often flattened, and even heart-shaped, in burrowing forms. An evolutionary process is evident when we study certain regular echinoids from the upper Carboniferous. In early forms, the ambulacral area consists of two rows of plates, each with a single row of pore pairs on each side, and therefore called **uniserial**. Gradually, more plates were added in the round forms to make biserial and multiserial arrangements. (See figure 17.) A second example of evolutionary change is that from radial to bilateral symmetry in regular echinoids. The test becomes heart-shaped in outline, and the ambulacral areas restricted to, so that they become petal-shaped on the apical surface. The peak of change is seen in *Micraster*. (See figure 18.)

The variety and healthy development of many different animal types reflects the favourable conditions under which they lived. The marine conditions had a stability quite unknown on land, nevertheless, the successive generations were far from being stationary in physical structure. Evolutionary trends are easy to follow in animals such as the *Echinoids,* with their armour-like framework. It is called a **test** not a 'shell', for it is interpenetrated by living substance. Interest in the structure of *Micraster*, the heart-shaped urchin, was aroused by the work of A. W. Rowe. In 1899, working in the limestone along the Kentish coast, he kept careful records of location, cleaned the specimens with great care, and subjected them to minute scrutiny. His work gave an idea of the echinoid evolutionary development. The test of an echinoid is a miniature, but wonderful, building made up of a multitude of calcareous plates, differing in shape and size, but beautifully fitted together. We can presume from the case study, that as the creature evolved, these plates travelled slowly outwards across the upper surface, descending beyond the **ambitus** or circumference. Having arrived on the lower surface, they moved slowly towards the mouth. Throughout this journey, the size, outline and curvature of each plate was being adjusted to fit in close accord with its neighbouring plates, and in harmony with the test as a whole. Eventually the genus *Micraster* revealed well-established bilateral symmetry. The mouth had shifted forward and the jaws had disappeared, the test was heart-shaped. (See figure 18.)

On the vast sea floor, with its uniform and stable conditions, its inexhaustible store of food, and its apparent lack of competition, the precise influences which stimulated serial changes must have been largely internal and genetic. No doubt they originated primarily as combinations of changes, or **mutations**, in the gene-complex, but their expression was guided by **selection**. From these and many other features in the *Micraster* study, there seems no evidence that the steps in these changes occurred at random.

What then is a mutation?

Is it the starting point of a new series, or is it each step in the sequence?

Can we accept the two concepts in evolutionary thought —'adaptive radiation' and the 'ladder of life'?

A satisfactory explanation for this apparent contradiction is one of the most pressing problems of evolutionary theory.

The Rise and Fall of Fossil Organisms

The fossil record seems to be marked, at times by great *proliferation of species,* at others, by *mass extinctions.* Sufficient is now known about the distribution of most fossil groups in successive rock strata around the world to expect answers to such questions as:

Is there a correlation in time between evolutionary spurts by different groups of animals?

Can the fluctuation in the flow of life be related to physical changes in the environment, reflected in the geological record?

Only a very tentative sequence of fossil zones can be suggested for the thirty million years of the Lower Cambrian, during which the

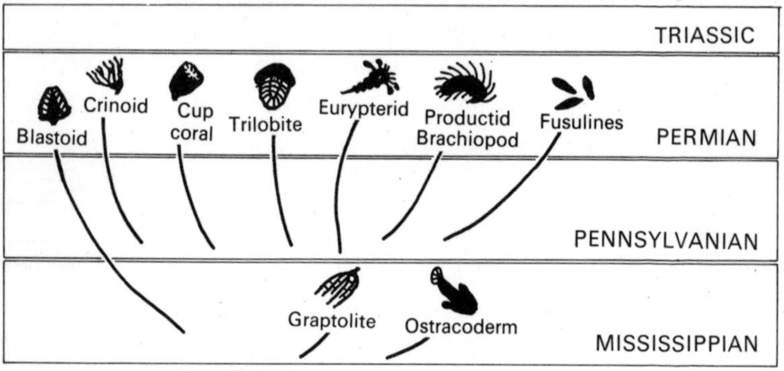

Figure 19A. Animals exterminated at or near the end of the Palaeozoic era.

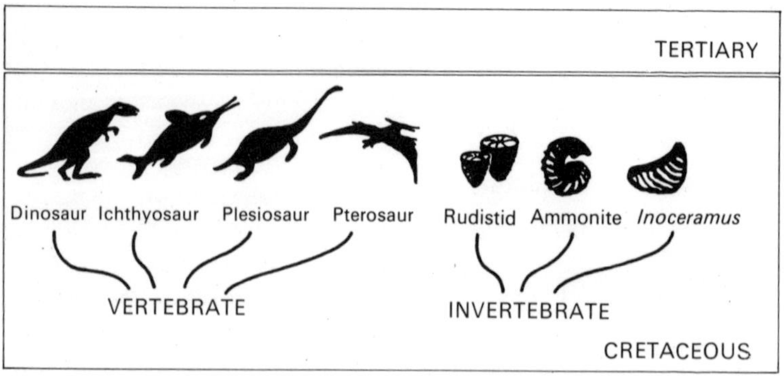

Figure 19B. Animals exterminated at or near the end of the Mesozoic era.

majority of animals developed hard parts, like shells. Whether similar, but soft-bodied, animals existed in the Precambrian, is not known. Perhaps the Lower Cambrian was a time of unusually rapid proliferation of marine organisms; or perhaps there was little expansion, and some environmental changes gave hard parts a selective value for the first time. Such changes must have had wide effects; they might include, for instance, the first appearance of carnivores.

A computer analysis by Funnell and Cutbill, Cambridge University, of the first and last appearances of fossil **taxa** (i.e., orders, families, genera, etc.) shows that the fluctuations in the distribution of taxa through time are not random. Some groups have double-peaked frequency distributions, often with a minimum in the Triassic period. Others are single-peaked, with maxima either before the Triassic (for example the trilobites) or later, as the group that includes star-fishes and sea-urchins.

In the Palaeozoic era, there was a rapid increase in the number of invertebrate taxa in the Ordovician period; it indicates an increase in the degree of exploitation of the Ordovician marine environments. There is still considereable discussion among palaeontologists about the geological evidence concerning causes of extinction. Although there are, of course, many possible causes, there are broadly two schools of thought:

- those who believe physical changes in the environment usually cause extinction, and
- those who consider that animal groups are commonly ousted by others.

Evidence supporting both points of view is available; perhaps both are right, in different situations!

Life in the Mesozoic Era

The Mesozoic was a time of transition and change in the plant kingdom. Late Cretaceous vegetation has little resemblance to Triassic flora. Seed ferns, lycopods and horsetails, all common in the Carboniferous coal forests, are represented by only small, insignificant species. Ferns, cycads, and cycad-like plants, now extinct, and the famous *Ginkgos,* (living fossils which may be the oldest living seed plant) were all present. Finally, in the mid-Cretaceous the flowering plants or angiosperms developed and set the scene for the diversity of this successful group in later periods. (See figure 20.)

The marine invertebrates made a slow recovery from the mysterious late Palaeozic extermination. Could it have been that the early Triassic seas were too cold, or otherwise inhospitable? There are virtually no coral reefs and very few beds of shells in Triassic rocks. What does this imply? Molluscs, successful cephalopods and echinoderms were represented by a variety of forms including starfish, sea urchins, crinoids and sea cucumbers. The majority of these organisms was capable of some degree of locomotion. Why must we consider this an advantage in survival?

The **Mesozoic** was a time of abundant land life, and is popularly known as the **Age of Reptiles**, because of the dominance of this

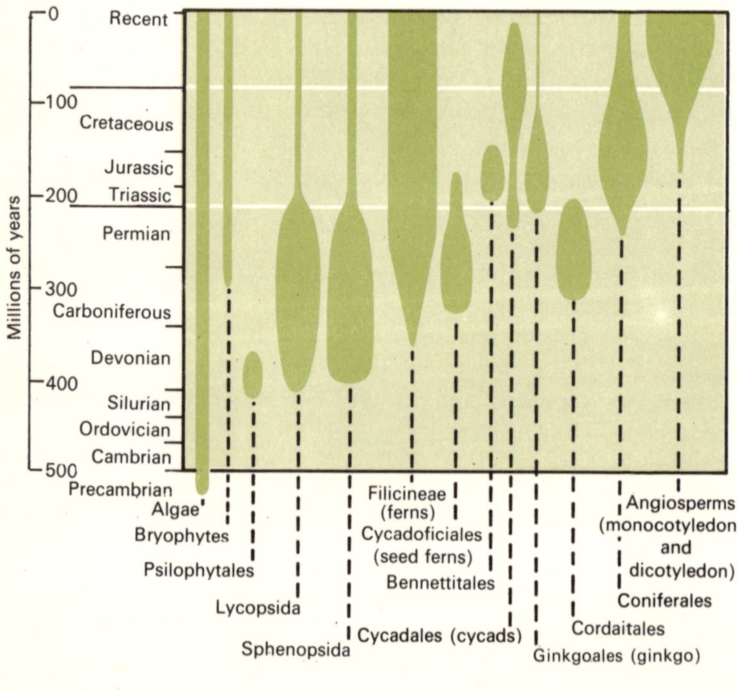

Figure 20. History of the Plant Kingdom.
Suggest what morphological trends may be associated with the emergence of these groups.
Discuss how the different groups are adapted to their environments.

Figure 21. Convergence on a grand scale.
These typical Australian mammals are all marsupials, only distantly related to the placental, or higher, mammals of other continents. Through independent evolutionary changes, the marsupials of Australia have given rise o dog-like forms, the Tasmanian 'wolf', and to others that are very similar to true mice, rats, squirrels, rabbits, and others

group on land, in the air and in the seas. Fish, amphibians and reptiles were dominant over 125 million years, compared with the 60 million years of the Cainozoic or Age of Mammals, in which man became superior. But it was in the Mesozoic that our modern forms have their origins via the mammal-like reptiles, the **therapsids**. The birds appeared during the late Jurassic, leaving their characteristics visible in well-preserved skeletons with imprints of feathers. (See figure 29.) The fossils of Mesozoic mammals are of great evolutionary importance although they are small, scattered, and almost insignificant in comparison with reptile remains. All are rodent-like, about the size of mice and rats, and several distinct orders are represented. They inhabited the undergrowth and branches of trees, ever on the alert for predatory reptiles with which they could not compete in open combat. Late in the period, the **marsupials** (pouched mammals) and **placentals** (advanced mammals) left their evidence in Mongolia and the western U.S.A.

The remains of hundreds of insect fossil forms have been found at Beacon Hill (New South Wales) and Ipswich (Queensland). They belong to 122 species, of 63 genera, 32 families, and 10 orders. This biologically successful class shows that diversity of form is a clue to survival. These fossils include wings with colour patterns, plant hoppers, cicadas, scorpion flies, ancient cockroaches, dragon flies and stone flies.

Life in the Cainozoic—the Era of Recent Life

The **Cainozoic era** (Greek: *kainos*-recent) commenced 63 million years ago and is the most recent, the shortest and the most complex to record.

Why should this be so? Because, although it is not unusually long, its physical and biological characteristics have been studied more thoroughly and are relatively more important than the geological histories of earlier intervals. It is the last great division of geological time, and we feel it is most important because it shows us the earth and its inhabitants as we know them.

You will see from table 7 that the era is divided into two periods and seven epochs. During this time there was a great multiplication in the number of mammalian fossils until, about the middle of the Cainozoic, there were organisms very like the modern mammals. In the youngest rocks there are fossils of organisms that closely resemble modern man.

During the **Tertiary period** few new major groups of plants and animals appeared, but many species evolved in response to changed conditions. Mammals were the dominant form of life on land. Birds and fish continued to evolve in their respective environments and became highly specialised. Plants migrated widely and there was a general withdrawal of warmth-loving vegetation towards the Equator.

Grass became the dominant vegetative form in temperate regions, and many raw materials, that later were to be important to human existence, were formed, things such as oil, gas and petroleum.

Glaciation was the characteristic of the later years of the Cainozoic, and resultant climatic changes were associated with factors such as the cooling of the oceans, the formation of fresh-water lakes in interior basins and the freezing of extensive tracts of soil and rock. Such changes caused great disturbance to the organic, living world, which must adjust to such non-living factors in the environment. Plants and animals were forced to migrate with changing conditions, and only the most suitable adapted to the new conditions. There were notable exterminations, especially among the larger mammals. It seems that the great mammoths and mastodons, 13 to 14 feet high at the shoulders, just lay down and died, to be preserved entirely in the ice sheets in northern Europe. The once-continuous plant populations became more widely separated as the climatic changes occurred and hence developed along specific lines to produce variety.

Evolution of the marsupial mammals had commenced in the **Cretaceous period** in North America and was followed by the appearance of man, a climactic event in the history of the Earth. It is certain that erect, two-legged primates had appeared in Africa by the early **Pleistocene,** and modern man evolved approximately 50 000 years ago. A vertebrate, a mammal and a primate—man, with superior mental powers, was placed in a unique position with respect to other forms of life—he could think about, and speculate upon his own origins!

History of the Vertebrates

The members of the *Vertebrata*, or vertebrate chordates as the largest and most advanced group within the *Phylum Chordata* are distinguished by the presence of a **vertebral** column, or backbone, composed of an internal living skeleton of **bone** or **cartilage**. Architecturally, this basic design offers certain advantages. One is the protection of essential and complex organs, and the possibility of having attached limbs that would facilitate more efficient physiological action and faster movement. For us who seek evidence in rocks of life on earth, such organisms leave a more complete and persistent fossil record, although vertebrate fossils are relatively rare.

The fossil history of animals with bone structure dates from the Ordovician, about 460 million years ago. As perhaps you may suspect, the fossils are incomplete and fragmentary, but as we have seen, the combined efforts of the specialists can help us reconstruct the characteristics of the early vertebrates, and relate present-day organisms to their ancestors. The study of fossil remains only, would leave us many problems in bridging gaps between fish, amphibians, reptiles, birds and mammals. The information yielded by the study of morphology, embryology, development and behaviour, show us that vertebrates are really a subphylum of the phylum Chordata, that seem to have some relationships in unsuspected ways. Apart from having a vertebral column that becomes evident in later life, the chordates, at some stage of their lives, possess a rodlike structure called a notochord. In most members, it disappears in adults, and is replaced functionally by the vertebral column of small bones. All the chordates possess a *hollow dorsal nerve tube* or spinal cord, and *slits in the side of the pharynx* or throat region during the early stages in development. These slits are retained as the obvious gills in fish. When such facts are established, it widens the field in which we seek similarities and pointers to development. So, we do not find it so

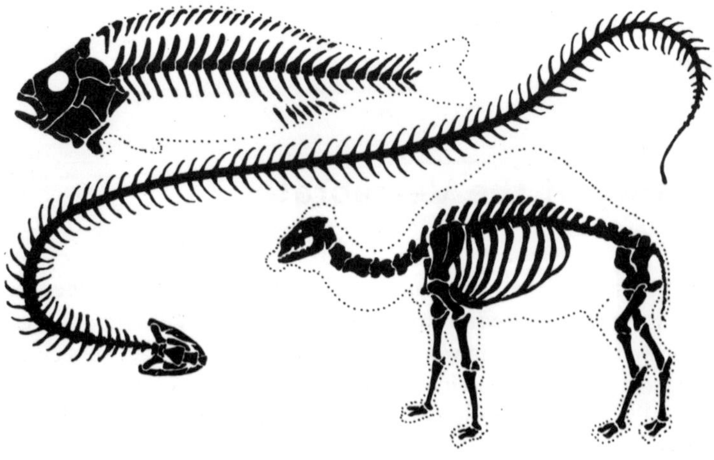

Figure 22. Variations on the vertebrate structure. What are some of the advantages of this variation?

incongruous that man and sea squirts are classified in the same phylum! If we now make a more detailed study of the evolution of vertebrates, we must go back through time to the Devonian period, 405 million years ago. (See figure 23.) The Devonian, you may remember was called the Age of Fishes.

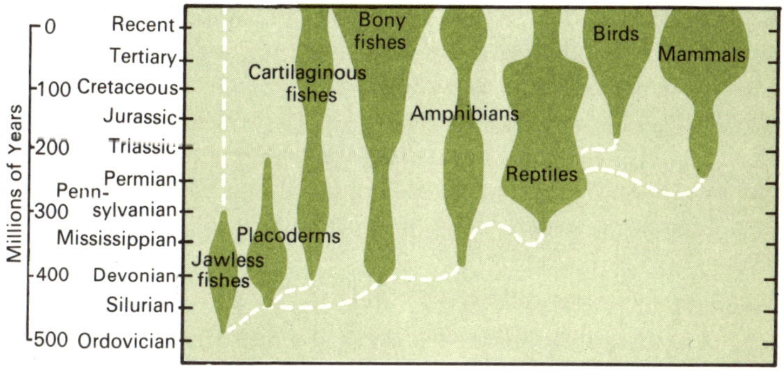

Figure 23. The historical record of the vertebrates. The width of the pathway is a measure of the relative abundance of the several classes. Can you account for differences in abundance at different periods?

The Age of Fishes

The first fishes were jawless **agnaths** (Greek: *a*-without + *gnathos*-jaw). They were primitive, few and unimpressive. Their appearance, however, is one of the most dramatic events in the history of life. It marked the rise of a phylum that was to become dominant in every sphere that it invaded, and was eventually to produce the human animals. The following facts are interesting:

The Ordovician sediments, in which the early scanty fossil record commences, indicate that the agnaths lived in fresh or brackish water. Called **ostracoderms**, they were jawless, armoured, bottom-dwelling fishes, with unpaired fins or without fins at all. The bony plating formed an outer skeleton while cartilage formed an inner skeleton. Which of these characteristics contributed to their name, ostracoderm?

They had a large number of gills. They are represented by the **hagfish** and **lamprey** today. The Devonian agnaths were erased from the ecological picture by another more efficient group of fishes, the **placoderms.**

Now, what would promote greater efficiency of life in the early oceans and rivers of the emerging earth? Perhaps a greater degree of symmetry and paired fins would facilitate movement. Perhaps development of jaws would permit a wider range of feeding, and hence permit existence in a wider range of habitats. The placoderms fitted the picture, having adapted to cope with the environment more efficiently than earlier forms. The placoderms appeared in the Silurian period to inhabit the extensive continental seas. They had bone in the external and internal skeleton, paired pectoral and pelvic fins, and upper and lower jaws with sharp searing blades. They were varied and highly successful through the Devonian period, but were extinct by the Carboniferous. Some appeared much like recent fishes, others were mud-feeders, and others violently predaceous.

While we would expect some of these variations to be 'biologically successful' it seems that nature often has to revert to a primitive basic plan and redevelop it for success at a different time and place. We find that the fishes that replaced the placoderms arose from early placoderms, just as the placoderms arose from early agnaths, and then replaced the later agnaths. Thus the replacing groups were finally successful, as the dominant fishes today arose in the smaller inland seas and colder oceans of the Devonian period.

These successful groups are the cartilaginous fishes. (Class **Chondrichthyes:** Greek: *chondros*-cartilage+*ichthyes*-fish.) They possess:

An internal skeleton made of cartilage. The absence of bone found in earlier groups seems a secondary development; gill slits opening independently to the outside.

Ancient forms were flattened with broad heads, and terminal mouths equipped with pointed teeth.

These fish were originally specialized for life in sea water, and today the group is still fundamentally marine. This requires adaptation to meet the problems of osmosis. Principal present-day chondrichthyes are the sharks and rays and skates that probably arose in the **Jurassic period.**

The bony fishes (Class **Osteichthyes:** Greek: *osteon*-bone + *ichthyes*-fish) appear to have developed from the same ancestors as cartilaginous fishes and we would expect these more developed organisms to have been the most successful of all the fishes, invading most of the waters of the earth. They were characterized by:

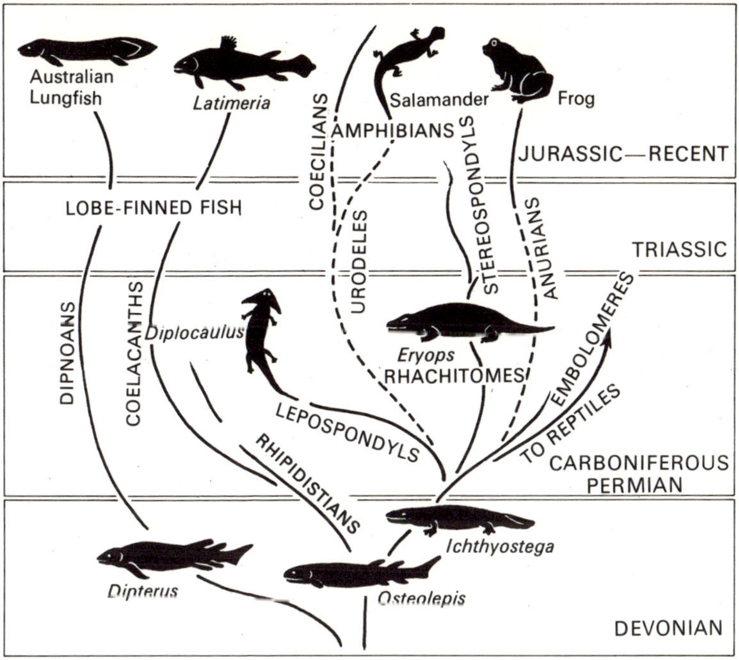

Figure 24. Evolution of the lobe-finned fishes and amphibians, showing live representatives and key fossil forms.

- the retention and intensification of bone in the skeleton, the earliest being heavily armoured and covered with bony scales:
- the gill slits, open in the Chondrichthyes, being covered by an operculum or gill cover.

Primitive Osteichthyes had lungs as well as gills, perhaps an adaptation to fresh water life or a supplementary means of respiration when oxygen became deficient. The osteichthyes, and especially the great group called **teleosts** ('perfect bones') are dominant aquatic animals today. Their diversity in form has developed in the long period from the Devonian, and more than 20 000 species are known. The abundance of a single species such as the herring in the Atlantic Ocean is astounding! One interesting group of the Osteichthyes was thought extinct until, in 1939, a specimen of the **crossopterygians** ('fringed fins') was caught off the South African shore. This discovery would be as extraordinary as finding a living dinosaur!

We sometimes suddenly find the past in the present when we investigate certain species such as the lungfish in South America, Africa and Australia and find that they are *living fossils*. Some of the primitive osteichthyes can be called air-breathing fish and give some clue about the manner in which living things eventually came to land. These fishes are very ancient, evolving early in the history of the bony fishes.

They had lungs and could breathe air when surfacing; therefore, they could exist in stagnant pools and swamps formed during the periods of great topographical change on the earth's surface.

Present-day observations show that the lungfish are river-dwellers, and can survive in water too low in oxygen to support fish that are entirely dependent on gills for gaseous exchange.

Another group of lobe-finned fish, the **rhipidistians**, now extinct, are considered the direct ancestors of the amphibians. It was thought

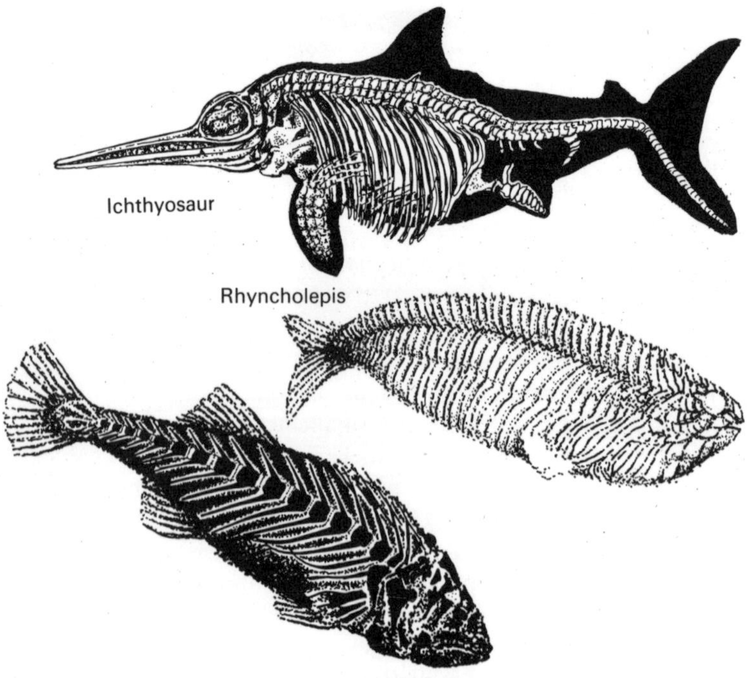

Figure 25. Fossils

that they could breathe air at the surface of the water, could probably crawl along the shores and even venture onto land to capture insects such as cockroaches and spiders, in abundance at the time. Can these rhipidistians be considered the water-land link in the earth's history? Fossil evidence suggests that the movement of life onto land was in full swing during the Devonian and was an accomplished fact by the end of that period. (See figure 25.)

Life on land—the First Arrivals

The late Permian, about 250 million years ago, was characterized by great changes in climate and topography. The levels of the continents rose all over the world, and a general folding of the Earth's crust produced mountain ranges, and hence, a number of terrestrial habitats. Swamps, highlands and arid lands were available for those life forms that could survive. And survive some of them did, although they were no longer surrounded by that necessary water, as their ancestors were. Great morphological and physiological changes were necessary for individuals to survive and reproduce the species. In an historical sequence of land occupation it seems logical that:

- Plants should have led the way, being the providers in ecological patterns. And this is just what happened.
- The smaller forms, such as primitive mosses, gave way to the larger, and an abundance of trees formed the first forest grown on earth.
- Land invertebrates, forming part of the world food chain, followed and ranged from arthropods to molluscs, the insects being the most striking and impressive in numbers and variety.
- The earliest land vertebrates appeared towards the end of the Devonian and were amphibians, a word derived from the Latin, meaning animals living in both elements.
- These animals apparently had the most successful combinations of variable characteristics, and survived to reproduce the species.

These animals, called **labyrinthodonts**, straddled from the water to land. It was not only lungs inherited from primitive fish that were the major land-oriented factor. It was also the change of paired fins to limbs.

In most other respects, the first amphibians were still completely fishlike.

So the late Palaeozoic era, the **Carboniferous period**, became the **Age of the Amphibians**. With their lungs, tough waterproof skins, strong skeletons and mobile pentadactyl limbs, the first land vertibrates had taken tentative steps on dry land! But they never became quite free of the need for abundant water, requiring it to reproduce and prevent absolute desiccation. Hence they did not spread over the entire land surface. They are represented today by frogs, toads and salamanders. Perhaps, in the end, their real claim to fame is that they were the forerunners of the reptiles.

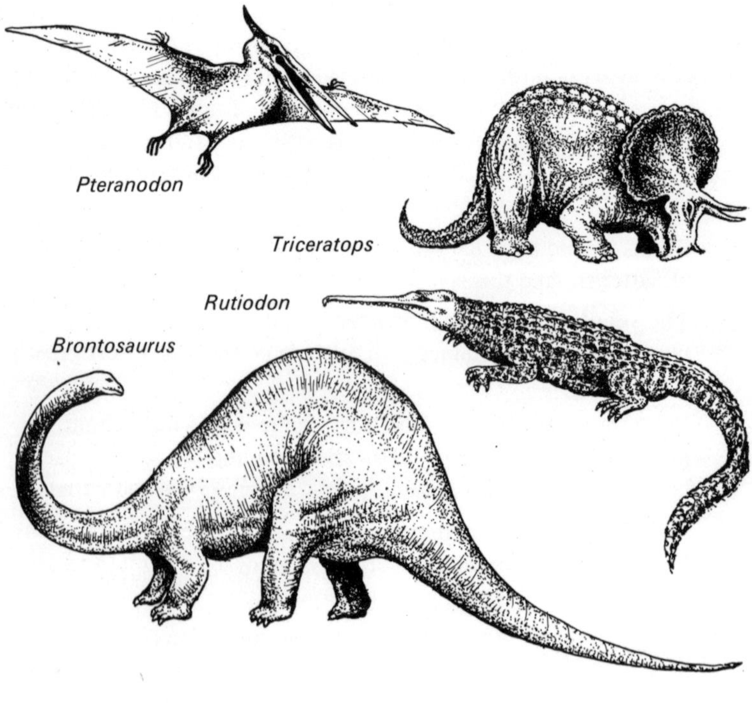

Figure 26. Some reptiles of the Mesozoic era. It is possible to deduce from their forms, the various habitats that they must have occupied.

The Age of Reptiles

The transition of the vertebrates to land life was completed when the reptiles emerged. The change was gradual and it is difficult to pinpoint its exact time, but some of the late Carboniferous fossils seem to represent reptiles. What was it that the amphibians lacked and the reptiles developed? An important evolutionary factor for success in a terrestrial environment is to be independent of water for reproductive processes. The reptiles reproduced successfully on land. Perhaps you can guess what this special adaptation was.

From the labyrinthodont amphibians emerged the primitive root or stem reptiles, the **cotylosaurs**, so called because they are the stock from which most, or all, the higher vertebrates branched and evolved. They integrated with earlier labyrinthodont amphibians and often looked not unlike those ancestors.

Into the Air

The only adaptation needed by terrestrially based animals to occupy the main unoccupied sphere was that of flight. The pterosaurs, 'wing reptiles' or pterodactyls, were the only reptiles to fly and they were not particularly efficient at this. They appeared in the early Jurassic and became extinct towards the close of the Mesozoic. With forelimbs modified as wings, these strange animals soared and glided but lacked the power of directional flight. They possibly lived near the Mesozoic seas, catching fish as they skimmed the surface. Inland land species do not seem to have existed.

On the Earth—the Time of Giants

Possibly the most familiar of all the fossil reptiles are those derived from the pelycosaurs, which earlier evolved from the cotylosaurs. The pelycosaurs gave rise to the **dinosaurs**, *'terrible reptiles'*, and amongst these was the dinosaur of science fiction, and actual fact, *Brontosaurus*. But this was only one variant amongst many. The dinosaurs came in a wide selection of shapes and sizes; some walked on hind legs, some on all fours; some ate meat and others, plants. (See figure 28.) There were: *Brontosaurus*, the 'gentle giant' weighing about 35 tons, *Tyrannosaurus*, the giant reptile standing 5½ metres tall, devouring with sabre-like teeth, and *Coelophysis*, running on long hind legs, a quiet meat-eater.

Figure 27. The dinosaur Tyrannosaurus rex.

Figure 28. Well preserved skull of Protoceratops *found in the Gobi Desert by a combined Soviet-Mongolian palaeontological expedition. You can clearly see joins between bones and openings for blood vessels and nerves.*

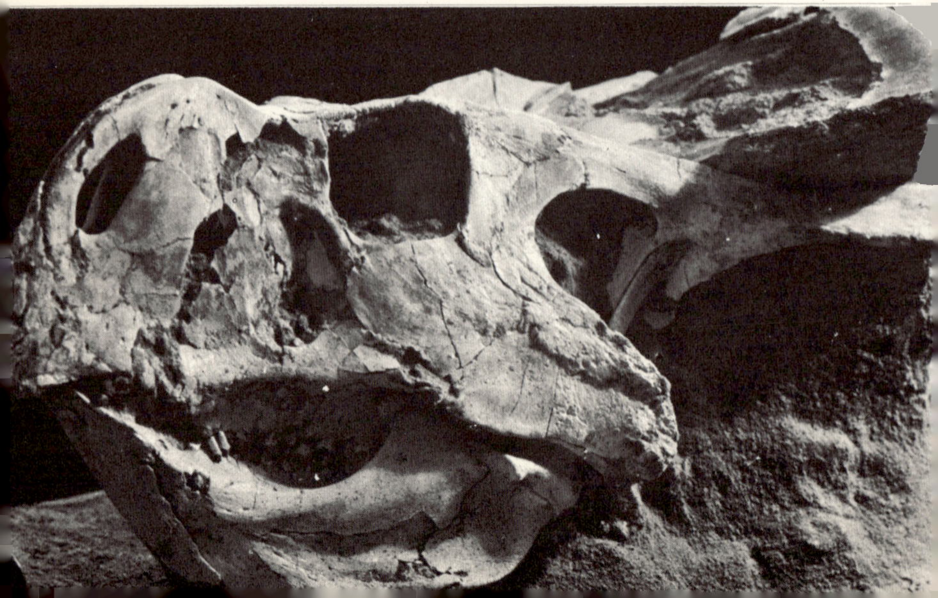

From the middle of the Permian to the Jurassic, great numbers of interesting reptiles developed, evolved from the pelycosaurs. One such, *Cynognathus*, the 'dog-jawed reptile', was a slender, lightly-built animal about 2 metres long, with limbs directly under the body, and teeth, not uniform as in reptiles, but differentiated into canines, incisors and molars. If we consider typical mammalian characteristics such as hairy skin, heart structure, blood temperature and the ability to care for the young, we see why there is great difficulty in obtaining fossil evidence for a transition from reptile to mammal. But perhaps animals like *Cynognathus* provide us with the link based on patterns of dentition and arrangement of limbs. These 'hopeful monsters' may have been the immediate ancestors of the mammals.

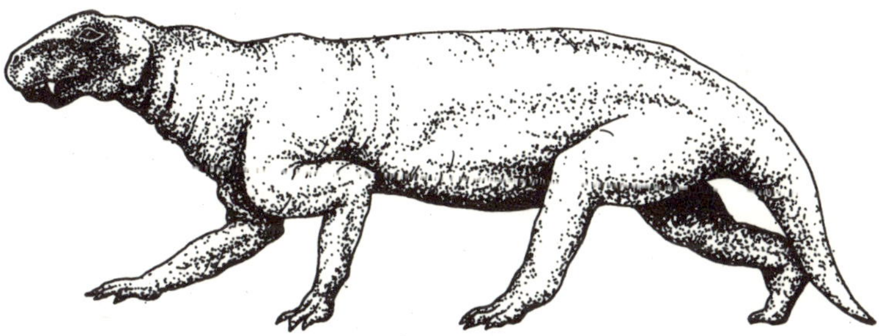

Figure 29. A therapsid—the first mammal? Cynognathus, *the 'dog-jawed' reptile.*

These primitive reptiles were eventually crowded out by a great variety of other reptiles, but not before they had provided an ancestral prototype for mammals. Of the Mesozoic reptiles, a few still survive, including lizards, snakes, crocodiles and turtles. But by the late Cretaceous period, only the skeletal remains were left, to tell us something of the great diversity of life in this period. The land-living dinosaurs and toothed birds became extinct, the great marine reptiles disappeared, leaving only a few remains, yet nothing befell the fishes on which they fed. The archaic mammals had arrived. How can this period of reptilian extinction be explained? It is only possible to hypothesize. Was there widespread environmental change?

But the reptiles occupied so many environmental niches, that surely some would have survived changed conditions. Other groups living

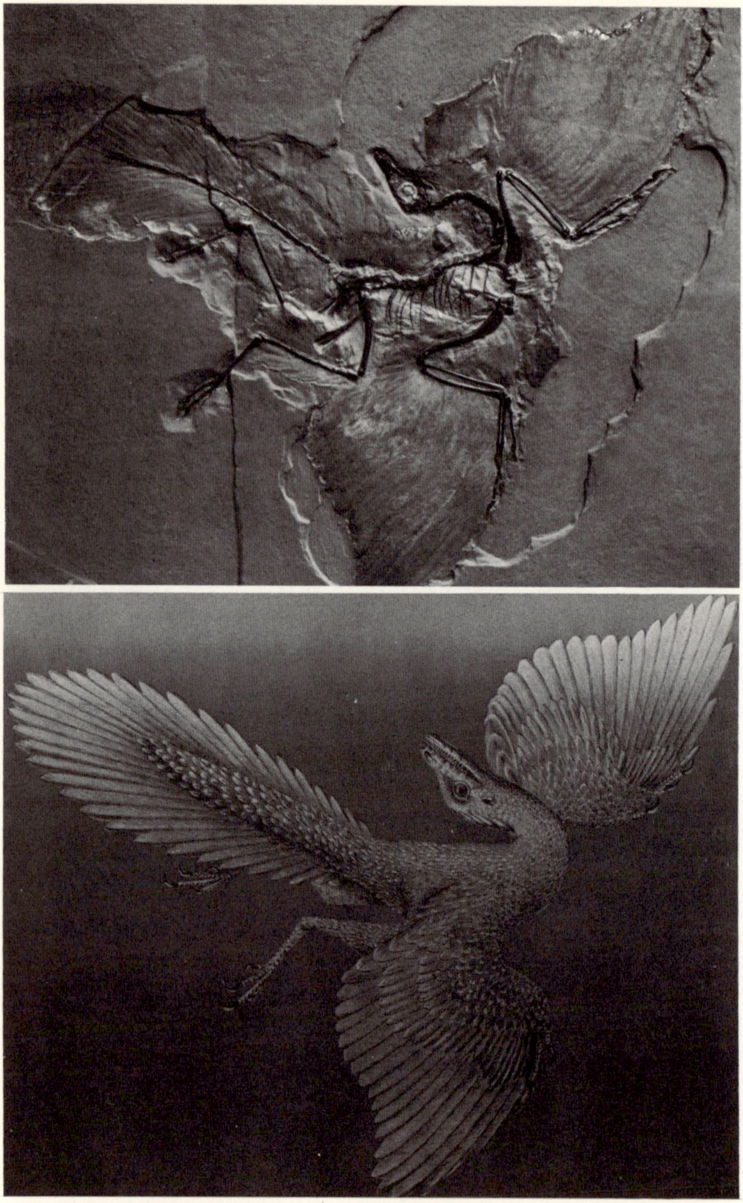

Figure 30. Archaeopteryx, *a key link in the history of vertebrates.*
(a) Cast of fossil Archaeopteryx.
(a) Restoration of Archaeopteryx.

with them did not die in these changed environments, and some did not even undergo evident change to cope with the new environment.

Could it have been only one factor that caused the *'great dying'* at the end of the Mesozoic era?

Many reasons have been advanced—even constipation resulting from changes in the available plant diet! However, the drop in temperature at the time is a likely factor, and it is suggested that while smaller reptiles survived, the dinosaurs perished because they were too big to find places to hibernate during the increasingly cold winters.

But a problem for this theory is that there was extinction of even small members of the dinosaur family. One was only about the size of a large turkey. The fossil evidence of these tiny forms may, of course, be incomplete and they could have survived until later times. So the story is confused. It does not seem logically necessary or probable that any one factor caused the extinction of the great reptiles. Even if it were, what factor could it have been?

The 'great dying' did occur. It closed an era and opened the world to a new era, the Cainozoic, the **Age of Mammals**. Now we must remember that fossils are facts, but relationships between the fossils are hypothetical. The pterosaurs, or 'winged reptiles', enjoyed a considerable but transient biological success, and disappeared in the 'great dying'. They were true reptiles and were not the ancestors of birds. A long way back in the early Mesozoic, at least 150 million years ago, perhaps some reptilian eggs hatched to forms that could no longer be considered reptilian, but were a new higher class, the birds.

Archaeopteryx, a 'flying reptile' which had feathers and such reptilian characteristics as teeth, a jointed tail, and bird-like claws at the extremities of the wings, is probably one of the most rare and famous of fossils. It shows an intermediate stage between true reptiles and true birds, and points to the direction which gave birds evolutionary success. Could this radical adaptive change have taken place in one mutational leap? Could, as the palaeontologist, Schindewolf, suggests, 'the first bird have hatched from a reptilian egg?'

Such a big leap is known as a **macroevolutionary change** and while it is usual to think that progressive development occurred by small steps or by *microevolutionary change*, it is really not so easy to rule out macroevolution. While the earth continues to keep evidence to herself and not yield a full fossil record, we have these gaps that can

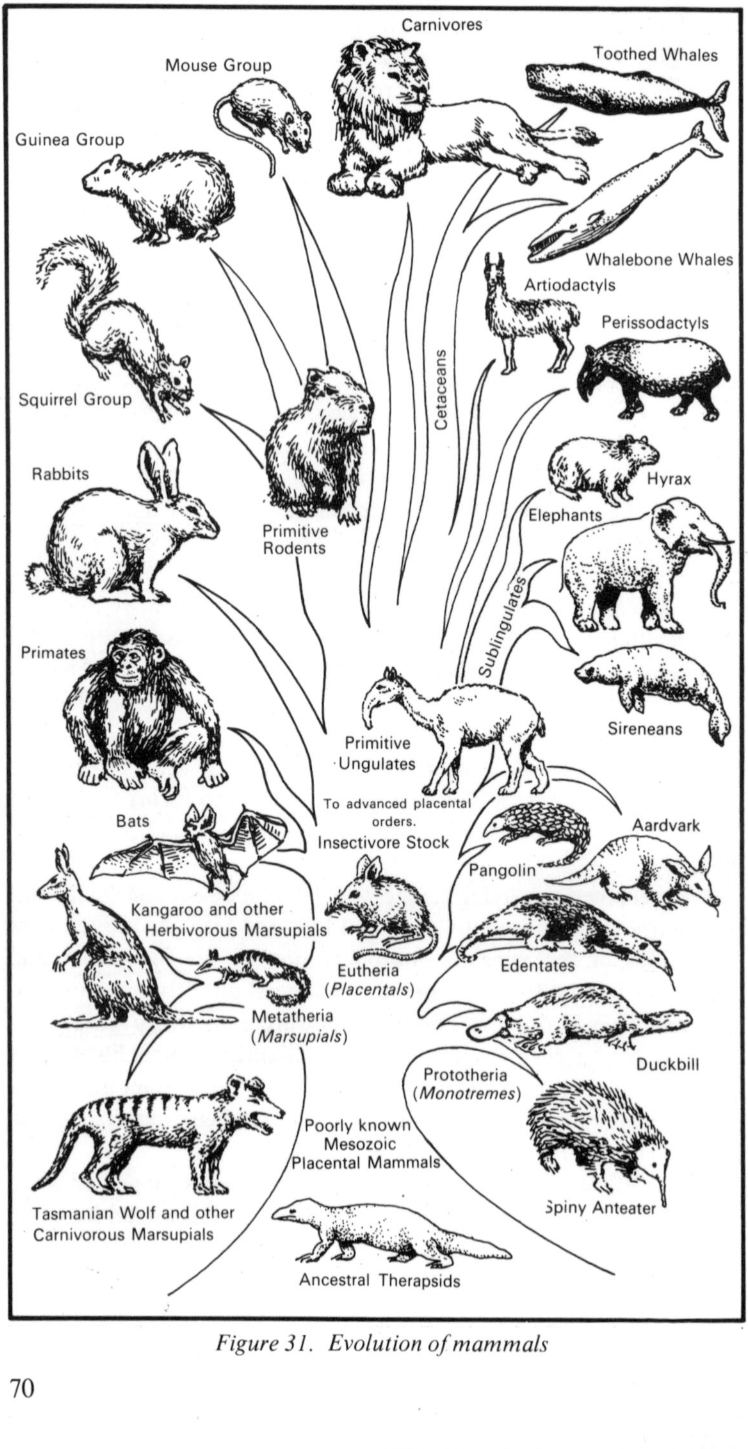

Figure 31. Evolution of mammals

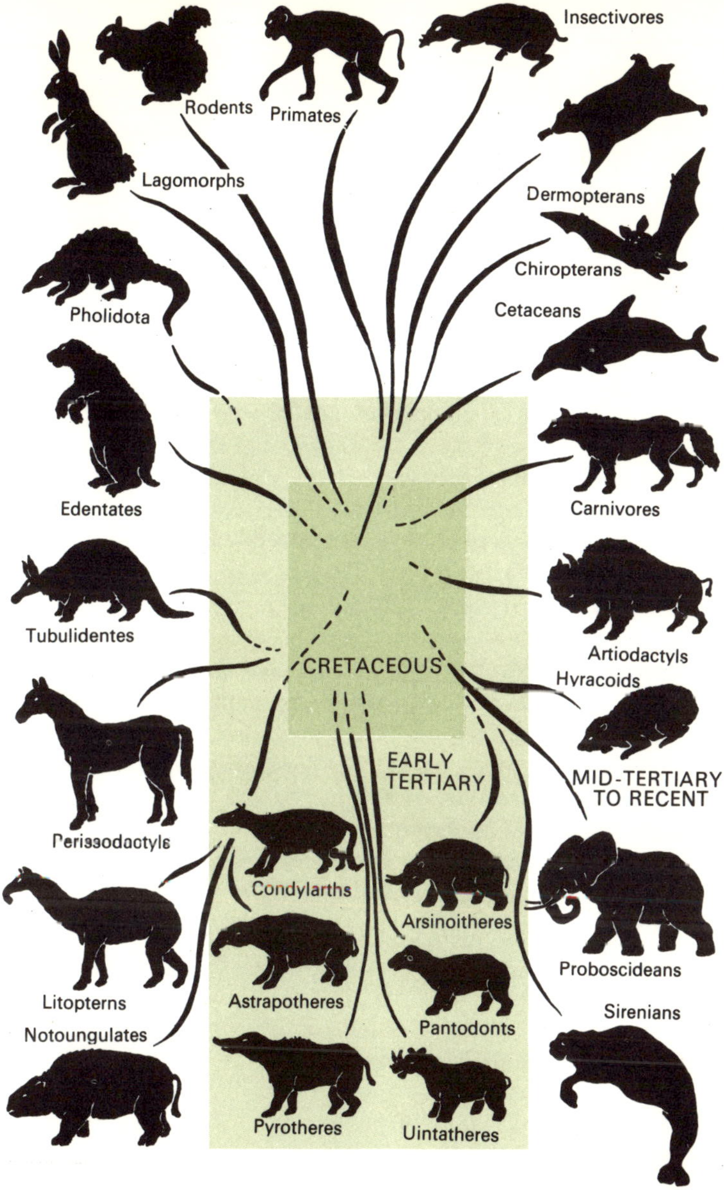

Figure 32. Radial adaptations and relationships of the various orders of placental mammals. Those forms within the square were extinct by the end of the early Tertiary.
Can you name a member of each order represented?
What principle underlies the arrangement of the illustrations in this diagram?

only be filled by hypothesis, or speculation. However birds came to be, they have, over the past 50 million years, successfully invaded almost all parts of the earth.

Evolution of Mammals and the Emergence of The Mind

The final, and to us, the supreme class of vertebrates, the **mammals**, arose from the reptiles (not with the appearance of any key characteristics but by a long-continued gradual change)—by microevolution. Appearing from mammal-like reptiles in the late Triassic or earliest Jurassic, for at least 75 million years, and perhaps as much as 90 million years, mammals cut a small figure in the world. That duration is at least as long as the whole era of their later dominance, the **Age of Mammals**. Mouse-sized and rare, it appears that some ate seeds and some fruits, some were predaceous and carnivorous, others were what is commonly called **insectivorous**, eating any small food bits of animal origin.

· Why were these animals so obscure for so long?

We can only hypothesize that amphibians had the scene almost to themselves. Reptiles, as we have seen, radiated adaptively and occupied many habitats at the expense of the amphibians. So, when the mammals eventually arrived, the earth was occupied by well adapted reptiles. There was a series of ecological **replacements**. The fact that mammals did survive at all must be attributed to inherent advantages not shared by the reptiles. Mammals possessed, of course, more efficient physiology and reproductive processes; brains did not help at this stage, for what evidence we have, strongly suggests that Mesozoic mammals gained no advantage over their reptilian cousins.

The Emergence of Mammals

The **Cainozoic** was a time of predominant, widespread uplift with consequent erosion, few deposits of sediments were made where we can get at them. As with much of the earth's history, scientists have put together only a patchy account.

Late Cretaceous rocks show dinosaurs in some numbers and also tiny mammals. The seed-eaters were to become extinct in the early Eocene, while insectivorous forms were mostly marsupials. Their secret of success was that they were warm-blooded (**homoiothermic**) and provided some care for the developing young by rearing them partly in a pouch.

In the earliest Palaeocene the dinosaurs disappeared with startling suddenness. Possum-like marsupials and placental insectivores were a small minority. Common mammals appeared larger, longer, rather short-legged, running on five toes, with tails long and heads small in proportion to their bodies.

By late Palaeocene, there was a rapid expansion and divergence of placental mammals—hoofed herbivores, predaceous carnivores of varied size and form, small forerunners of the monkeys, rare rodents and others. The mammals were finally inheriting the earth. It seems that the 'great dying' had left the empty environments to the mammals, who occupied them in a world-wide adaptive radiation most active during the Palaeocene. (See figure 32.)

More efficient, and more capable of adaptation, the mammals went further than the reptiles ever had in occupying environmental habitats. All this only makes more mysterious and exasperating the problem of why the Mesozoic reptiles became extinct. It does not seem that they became extinct because the mammals had increased in number and variety. In fact, the mammals seem to have increased because the reptiles became extinct.

But why?

Can you suggest ways in which man may eventually find the answer?

Modernization of the Mammals

The Eocene was the link of the old with the new—the modern orders of mammals were arising and the older ones were not yet extinct. Indeed, there were more orders of placental mammals then than there have ever been since!

Ancient and now extinct groups of mammals, prominent in the Eocene and Oligocene, included the **titanotheres**, heavy, belligerent looking animals rather like the modern rhinoceros. The now-extinct carnivores such as the sabre-tooths were replaced by the bulky probiscidians, such as the elephants and their allies. Using the techniques available, G. G. Simpson, in 1951, published an interesting account of the horse's ancestry, including the complicated tree of horse cousins which branched off and are now extinct. Of course, not every family tree is as well worked out as that of horses. In fact, the many higher forms of plants and insects, for example, seem to have developed so quickly that no disentangling of ancestry is possible at present. As we near the era of man, we may forgive ourselves for being more interested in our own history than that of the horse.

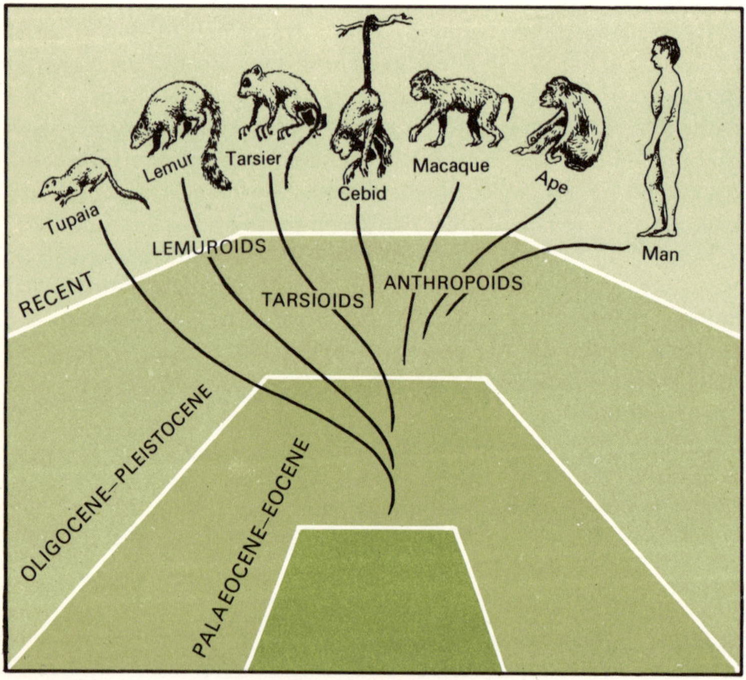

Figure 33. Relationships and inferred evolution of the major primate groups. Notice that the order of appearance, as indicated by the length of the family lines, is also the order of zoological ranking. What does this imply?

The fossil record of the primates (Latin: *primus*-first) is, in general, quite extensive, but some gaps will always remain because the fossils are simply not preserved. (See figure 33.) At this time, the history of the primates was more far-flung, with crucial episodes in many different regions. Sometimes these occurred in tropical areas which are, as a rule, neither very rich in fossil deposits, nor thoroughly explored for them. The primate way of life also precluded a good fossil record, for most are, or have been, arboreal, and therefore are comparatively unlikely to be buried or to fossilize. And perhaps, the very element of shrewdness that facilitates survival also works against the possibility of fossilization. Let us survey the earth now:

> 'The placentals having arrived, they are dominant land vertebrates; divided into twenty seven orders, of which, however, only sixteen have living representatives, their abundance is evidence of the remarkable adaptive radiation that has occurred among placentals in the last 70 million years or so; shrews and moles, bats, armadillos, rabbits, squirrels, porcupines, mice, whales, cats and dogs, elephants, horses, pigs, camels, sheep and cows. The placentals burrow, fly, climb, and run and swim; they eat worms, fruit, insects, grass, seaweeds, squids, crustaceans, bark, cocktail savouries and each other. They live in the open sea, in tropical treetops, on Arctic ice floes, in affluence and poverty, and in sandy deserts'. (*Simpson et. al. Life.*)

The Primates and the Emergence of Man

Refer to NSCM Booklet B19 Evolution of Man I, for a more detailed account of the primates and of man's evolution.

The order of primates arose from amongst the most primitive of placentals, the early insectivores, and seems, at first to have little to distinguish it, beyond the use of the forefeet as hands and increasing co-ordination of visual perception and manual response. These primitive creatures, the lemurs, tarsiers and tree-shrews, had the potential to evolve a high degree of intelligence. This they did; we are not completely certain why or how it occurred.

Among the early primates, which were abundant and lived all over the world except Australia and South America, were the **prosimians,** the lemurs and tarsiers, who have shown a micro-radiation in isolated areas in Madagascar and Africa. From these evolved the **New World monkeys** represented by the **African monkeys** as rhesus, baboons and mandrills; and the climax group, the **Hominoids**, including gibbons, apes, chimpanzees and man.

Some of the features that have permitted emergence of the primates as the dominant group on the earth are: binocular vision, that helps in judging distance and perspective; specialised teeth, permitting varied diets; complex nervous organs (such as the brain, making possible reception and interpretation of data in a competitive world); longer limbs with hands, facilitating grasp and helping interpretation of touch sensation; and increasingly larger bodies, giving dominance over other organisms. Variations upon these basic features make them suitable for every environment on the earth's surface; their adaptive radiation is extensive.

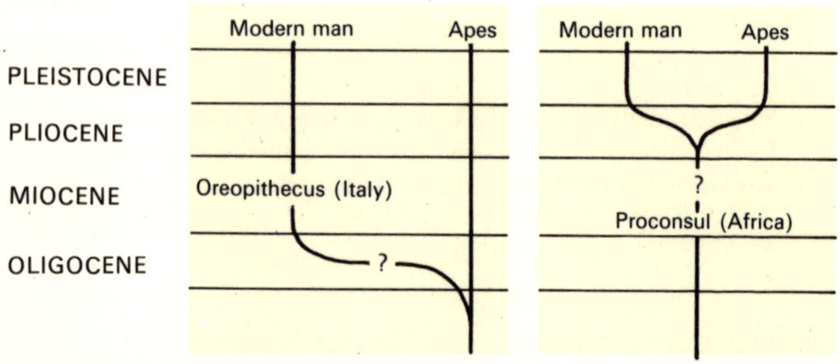

Figure 34. Two possible origins of man.
If Oreopithecus *is the ancestor of modern man, the point of departure must have been in the Oligocene; perhaps before ape-like characteristics were fully established.*
If Proconsul *is an ancestor of modern man and modern apes, then the distinctly human stock began later; perhaps in the Miocene or Pliocene.*

From the Transvaal in South Africa came the evidence that man's origins lay in early stock of **anthropoid apes**. It seems that the ape-like hominid **Australopithecus** walked upright, and existed as recently as one million years ago. They interpreted their environment with larger brains than apes but were not our ancestors, for one million years is rather too short a time for modern man to have emerged. *Australopithecus* provides evidence for the kinds of changes that must have occurred for the common ancestor to have given rise to man.

Perhaps the fossil remains of **Ramapithecus**, found by Dr L. S. Leakey, in Kenya, points the way to our own history. About 14 million years old, the jaws and teeth are more like man's than apes, and probably *Ramapithecus* was the first of our ancestors recognisably different from the main stock of apes. However, he could not be called '*man*'.

Where do we find the strongest clue to our origin?

The evidence comes from links with the common ancestor of apes and man, from widely scattered parts of Africa, Asia and Europe and from reports of archaeologists who have found and described the 'fossil men'. *Homo erectus* stood upright and was about five feet tall, with half-size brain and primitive culture. *Homo neanderthalensis* had a brain the same size as modern man and appears to have been intelligent and resourceful. He was a side branch to the eventual *Homo sapiens*, so recent that his history dates from Cromagnon man, some 40 000 years ago. A fuller treatment of this fascinating topic will be found in the NSCM Booklet B20 The Evolution of Man Part II.

Efforts to determine the course of evolution and the history of living things do not have to end at this point, or continue to be restricted solely to fossil evidence.

A new discipline, **chemical palaeogenetics**, concerns the exact relation and order of origin of the major groups of organisms, the evolution of the genetic and metabolic complexity of present day organisms, and the nature of biochemical processes.

The millions of proteins existing today are, in effect, living fossils; they contain information about their own origin and history and about the ancestry and evolution of the organisms in which they are found. Computer analysis of protein evolution will give information to construct the history of living things in detail.

So, you see, we have come the full circle in our enquiry into the history of living things; the history of life lies in the secrets of molecular biology; much more of it will be known before the end of this century.

> 'It is the true office of history to represent the events themselves, together with the counsels, and to leave the observations and conclusions thereupon to the liberty and faculty of everyone's judgement'.
>
> Francis Bacon. A.D. 1561-1626

Reading List

Colbert, E. H. 'The Ancestors of Mammals'. *Scientific American* Offprint 806. This deals with therapsids and other mammal-like reptiles.

Dayhoff, Margaret Oakley. 'Computer Analysis of Protein Evolution', *Scientific American,* July 1969. This article tells of frontier research. It suggests that amino acid sequences of similar proteins in different organisms contain information on relations among species. This information is analysed, to reconstruct the history of living things in detail.

Eglinton, Geoffrey, and Calvin, Melvin. 'Chemical Fossils', *Scientific American*, January 1967. These research scientists tell how they have investigated certain rocks three billion years old, and studied the organic compounds they contain.

Ehrensvard, Gosta. *Life: Origin and Development*. Univ. of Chicago Press, 1961. The author is professor of biochemistry at Lund University, Sweden. He presents a view of the general trend of mechanisms behind the development of marine and terrestrial life. There is no scientific evidence on some of the theories outlined, but they are based on present knowledge of biochemistry, geochemistry, palaeontology, zoology, and botany, and are, at least, educated guesses.

Life Nature Library. *Evolution*. 1964. Time-Life International. A well-illustrated survey of many of the matters concerned with life history.

McAlester, A. Lee. *The History of Life*. Foundations of Earth Science Series, Prentice-Hall, 1968. Designed to permit students to share in the excitement of modern research, this survey is readable, and an up-to-date introduction to this aspect of earth science.

Millot, J. 1955. 'The Coelocanth'. *Scientific American* Offprint 831. Includes a brief account of the evolution of fish, a good reference chart and an interesting discussion of the coelocanth.

Stokes, W. L. *Essentials of Earth History*. Prentice Hall, 1966. A very good survey of geological and evolutionary processes.

Swinnerton, Henry Hurd. *Fossils*. New Naturalist Series. 1962 London. Collins. With a broad academic background, Professor

Swinnerton is not a narrow specialist. This work is a fascinating guide to the life of the past, and introduces the work of men and women who uncovered the secrets. It is a human as well as scientific story.

Wolterek, Heinz. *What Science knows about Life; an Exploration of Life Sources*. M.I.T. Press, 1965. A highly readable introduction to modern biology. Basic questions concerning the origins of life, the evolution of man and the immortality of living creatures, are discussed in the light of recent advances in biological research.

Revision Questions

1 What resources do scientists have for investigating past events on the surface of the earth?
Do you consider that there is sufficient evidence from which to write a history of living things? Support your decision, using examples.

2 Briefly outline two hypotheses that attempt to explain the origin of life on earth. Which do you favour and why?

3 Discuss reasons to explain the incompleteness of the fossil record.

4 Test your knowledge of the geologic time scale. Are you able to name the major periods, the length of their duration and the dominant life forms existent during those periods?

5 What advantages and disadvantages do you see in using case studies to develop a history of living things? Support your argument with reference to a specific case study.

Discussion Questions

1 Museums, such as the Australian Museum, in Sydney, reconstruct fossils as drawings or models. These reconstructions show complete organisms, yet frequently the fossil evidence from which they are made is fragmentary. What justification is there for doing this?

2 An article in *New Scientist* (6 Jan 1966) by Dr W. S. McKerrow is entitled 'The Rise and Fall of Fossil Organisms'. It is a report from the Geological Society and Palaeontological Association of Great Britain and reviews present knowledge about such fluctuations of life forms and their possible causes. What technological advances are helping provide data for such reports?

3 Imagine living 100 million years hence.
What dominant fossil forms would you expect to find and how would you deduce the nature of the environments and habitats in which the organisms lived?

Index